Dante and the Early Astronomer

Dante and the Early Astronomer

Science, Adventure, and a Victorian Woman Who Opened the Heavens

Tracy Daugherty

Yale UNIVERSITY PRESS New Haven and London

Published with assistance from the Louis Stern
Memorial Fund.

Yale University Press books may be purchased in
quantity for educational, business, or promotional use.
For information, please e-mail sales.press@yale.edu
(U.S. office) or sales@yaleup.co.uk (U.K. office).

Set in Caslon type by IDS Infotech, Ltd.
Printed in the United States of America.

Library of Congress Control Number: 2018959330
ISBN 978-0-300-23989-8 (hardcover : alk. paper)

A catalogue record for this book is available from the
British Library.

This paper meets the requirements of ANSI/NISO
Z39.48-1992 (Permanence of Paper).

10 9 8 7 6 5 4 3 2 1

For the due ragazze,
Hannah and Anna

Mary ran with haste to the hilltop country.
—Dante, *Purgatorio* XVIII

. . . an ancient poet seems to wait for us, smiling, on the crest of the modern cosmos.
—William Egginton

Contents

Preface
The Dawn-Light of Ravenna

In the pleasant calm of a clear spring morning in Ravenna, Italy, it can take a few minutes to awaken fully into consciousness—perhaps as many minutes as it takes light from the sun to reach the tops of the pine trees ringing Ravenna's gardens. There, mourning doves and magpies sing to their young, and there too, centuries ago, Dante Alighieri wrote the final lines of *The Divine Comedy*'s third *cantica*, *Paradiso*, proposing an entirely new universe.

Three centuries after Dante, in the early decades of the 1600s, his countryman Galileo Galilei replaced Dante's Aristotelian-Ptolemaic cosmos. Three hundred years after that, in November 1915, Albert Einstein—like Dante and Galileo the harried victim of an arrogant nation-state fearful of his ideas—introduced his general theory of relativity at the Prussian Academy of Sciences and appeared to coincidentally echo major aspects of Dante's astronomical vision.

The story of these three men is, in part, a story of how long it takes the light of understanding to reach us. It is also the story of how an individual's mind can hold the spectacle of infinity. Each man's life has been well documented, but few people have grappled with their scientific concepts as well as did a woman who lived on a hilltop in southern India in the first quarter of the twentieth century. She was in a prime position to appreciate their work. For her husband and her colleagues, she became a magnet around which fascinating ideas clustered. Mary Evershed was her name. What follows, in these pages, is her intellectual history. More surely than any of her peers, she traced Dante's universe, grasped Galileo's use of Dante's art to advance his scientific notions, and participated in the mental revolution that expanded existence the way Einstein—and Dante—predicted it would appear.

Late in his life, Galileo informed a friend by letter that he had gone "irreparably and completely blind; in such a way that that sky, that world and that universe, which with my wondrous observations and clear demonstrations I amplified . . . , is for me now so diminished and narrowed that it is no greater than what my body occupies."[1]

Beneath this lament we register the miracle that Galileo celebrates: how a weak human body, its frail mind doomed to erosion and death, can still gaze upon "that sky" and amplify the universe. This is precisely what Dante did in *his* last, failing days in Ravenna, watching dawn rouse the birds of Paradise; what Einstein did, risking his life in Berlin's anarchic streets to contemplate sunlight's path to Earth—and what Mary Evershed did on her Indian hilltop, deciphering the works of these three adventurers while observing sunspots, eclipses, and solar prominences.

My passion for the writings of Dante, particularly *The Divine Comedy*, is widely shared among contemporary poets and literary scholars. Every year new English translations of the *Comedy* appear (including, in 2004, a version by Sandow Birk and Marcus Sanders, and illustrated by Birk such that Hell is filled with ATMs and McDonald's french fries).[2] Eminent poets such as W. S. Merwin and Edward Hirsch have testified that Dante feels to them very much like a peer, the themes of his work remaining inexhaustible. Hirsch cites the "radical audacity of Dante's imagination" and the "seriousness and gravity of his epic task." Although "Dante as a national bard, a canonical figure whose work is recited by schoolchildren and engraved on Italian monuments, leaves me untouched," Hirsch says, "I have been overwhelmingly affected by Dante's story of himself descending into the bowels of the earth . . . and emerging, like Job, to tell the story of his strange journey."[3]

Dante can be read as a religious poet or a political poet—or he can be read in nonideological terms, as a marvelous observer of human behavior and as a master portraitist. His infatuation with Beatrice Portinari remains a favorite romantic tale among readers, as evidenced by Harriet Rubin's popular treatment of it in *Dante in Love* in 2004.[4]

In recent years, I have been excited to learn that Dante charms scientists as much as he does literary practitioners. Writers such as the mathematician Mark A. Peterson, the science historian Edward Grant, and the historian William Egginton have argued persuasively that Dante's medieval concepts of the shape of the universe strike contemporary astronomers as astonishingly prescient.[5] "At some deep level, poetry and physics are similar endeavors," Peterson writes. "Dante's universe . . . is almost unbelievably apt and accurate."[6] I became curious as to why and how this came to be the case, a curiosity kindled by the simple desire to understand more clearly the specific astronomical references in *The Divine Comedy*.

Some years ago, my pursuit led me to an author named M. A. Orr. I learned that she had published a useful reference book in 1913 called *Dante and the Early Astronomers*. I ordered an out-of-print copy through my local bookstore. The book proved to be a marvelous concoction of poetry and science, and I found myself as intrigued by the author as I was by the book's erudition. The edition I had found included a five-page introduction by the Dante scholar Barbara Reynolds. She revealed little about M. A. Orr—just enough to stir my imagination. Who *was* this woman, a late-Victorian-era lady sitting on a remote mountain outpost in India writing about Dante as though he were an astronomer?

This question set me on a decade-long journey to find traces of M. A. Orr, later known as Mary Evershed. By locating distant relatives in rural England and corresponding with archivists at the Science Museum in Wroughton, Wiltshire; the Royal Astronomical Society in London; and the Kodaikanal Solar Observatory in India, I eventually managed to piece together a portrait of this remarkable person.

Like Dante, Mary is a figure from the past whose story retains an extraordinarily contemporary flavor. As a woman excluded from fulfilling many of her professional ambitions because of cultural biases, and particularly as a woman trying to make her way in the sciences, she speaks to the experience of many women today.[7]

Mary's life interweaves political history, cultural history, and the history of ideas; it illuminates the cross-fertilization of literature and science. Her story offers a fascinating glimpse into the world of professional astronomy just before Einstein's theories revolutionized our view of the universe. It provides a unique perspective on issues such as colonialism, modernization, and globalization, and it offers the adventure of romance and travel as well. My modest hope is that readers will be as enchanted with Mary as I am, and that you will find her story as relevant today as Dante's is.

Acknowledgments

An Oregon State University Center for the Humanities Fellowship in 2010 helped me get this project off the ground. I am grateful to my former colleagues David Robinson, Wendy Madar, John Byrne, and the late Peter Copek.

Terry and Bryan Evershed have been unfailingly helpful, providing family background, information, photographs, and documents to fill in the story of John and Mary.

Ed Cliver of the National Solar Observatory and Nat Gopalswamy, a former director of the Kodaikanal Solar Observatory, now with the Solar Physics Observatory at Greenbelt, Maryland, kindly facilitated introductions and contacts for me.

For their kindness in responding to my inquiries, and for their help, research skills, and insights, I am grateful to the following: at the Indian Institute of Astrophysics, S. Sreekumar and Christina Birdie; at the Kodaikanal Solar Observatory, R. Ramesh, P. Prabahar, and Ebenezer Challasamy; at the Science Museum at Wroughton, Amanda John and Jasmine Rodgers; at the Royal Astronomical Society, Robert Smith, Helen Weeks, and Sian Prosser; at the Special Collections division of the University Library at the University of California at Santa Cruz, Luisa Haddad.

Colleen Mohyde took a chance and made it all happen.

At Yale University Press, I am most grateful to Joseph Calamia for his enthusiasm, kindness, and keen editorial skills. I am indebted to the Yale team, including Eva Skewes, Ann-Marie Imbornoni, Sonia Shannon, Nancy Ovedovitz, and Katie Golden. Anne Canright did a superb and very thoughtful copyediting job.

Thanks to my friends at *The Writer's Chronicle* and the University of Georgia Press. John Griswold and Supriya Bhatnagar offered useful comments on early versions of some of this material.

Brandon Brown gave me sage sentence-level advice.

My childhood pal Pat Edmiston was the first physicist I ever knew. He instilled in me a love of the West Texas sky and a passion for inquiry. Tim Lodle and David Stall continued his guidance years later, on many meteor-filled nights at our makeshift observatory just outside Midland, Texas.

Ted Leeson and Betty Campbell, Deborah and Creighton Lindsey, Keith Scribner and Jen Richter, Molly Brown, George Estreich, Sue and Larry Rodgers, David Turkel and Elena Passarello, Karen Holmberg and Aria Minu-Sepehr, David Biespiel and Wendy Willis, Martha Low, Jay Clarke, Scott Nadelson and Alexandra Opie, Jon Ross and Larisa Zimmerman, Terrance Millet and Denise Magee, Chick Gerke and Robin Strauss, and Bob and Mary Jo Nye provided sustaining friendship during the writing of this book. My love to Debra, Charlie, and Joey Vetter, and to Arlo Mullin.

Tom Stroik and Michelle Boisseau took my wife and me to the hills above Florence, where we gazed each day on Dante's beloved Baptistry. They were marvelous companions, along with our daughters, Anna Boisseau and Hannah Mullin. Michelle is now with Dante in the stars. With Jon Sandor and Maryann Wasiolek, my wife and I shared the loveliness of mornings in Ravenna.

Always, my love and devotion to Marjorie Sandor, the center of the spinning wheel.

Dante and the Early Astronomer

1

On the Hilltop

In 1920, T. S. Eliot complained that the astronomical "scaffold" in Dante's *Divine Comedy* was "almost unintelligible."[1] Eliot saw Dante's stars as part of an allegorical "presence" that guided readers to the *Comedy*'s emotional core and "into the divine." To appreciate the poem, Eliot wrote, it is not necessary to measure Dante's universe; one need only glimpse it, the way one catches faint starlight by glancing away from the source.

Ten years before Eliot aired his complaint, a self-educated amateur astronomer named Mary Acworth Evershed sat on a hill in southern India, seven thousand feet above sea level, staring at the daylight moon as she grappled with apparent mistakes in Dante's cosmos. The Aristotelian-Ptolemaic structure on which Dante based his universe had long been discredited, but that wasn't the sort of error absorbing her. Precision was her concern. Further, she wanted to enter Dante's literary imagination, to determine if he was accurate *within* his conception. Was he, as Eliot would claim, "unintelligible"? Or was he, for a man of his time and place, as insightful as one could be about the sky?

The moon tugged at Mary Evershed and stirred the twin tides of her passions: poetry and science. Dante was the poet who most thrilled her. As a longtime stargazer and the wife of an observatory official, she couldn't just skim Dante's cosmography. The *Comedy* mentions stars fifty-five times, with numerous other references to the moon and planets, to the constellations, and to seasonal measurements. As she watched the day-moon set, she recalled Dante's little-known Latin treatise the "Quaestio de Aqua et Terra," in which he appeared to make a troubling lunar blunder.[2]

In the "Quaestio," Dante argues that the moon is always in perigee—that is, at its closest approach to Earth—near the Earth's southern hemisphere. (Because of its elliptical orbit, the moon's distance from Earth varies between about 221,829 miles and 252,898 miles. From our terrestrial vantage point, the moon appears to travel eastward against the backdrop of night stars at a rather fast clip, approximately thirteen degrees in twenty-four hours.) Despite Dante's argument, medieval scientists knew the moon's perigee shifted both north *and* south of the equator, because the path of its orbit routed its journey east near all twelve of the constellations composing the zodiac. Dante's assertion seemed sloppy—in which case the *Comedy*'s scaffolding might very well contain loose steps.

(His slip appears in a larger treatise that we today know to be scientifically unfounded, but that is beside the point: *fantastical precision within a clear framework*, even if the details might be disproved, is part of Dante's charm.)

The "Quaestio" insists that land exists only in the Earth's northern hemisphere, with the exception of the island-mountain Purgatory in the south. Dante wonders how this could have occurred, geologically. Did the moon's "elevational influence" wrench mountains out of the northern ocean?[3] Certainly not, Dante writes. Everyone knows the moon's "influence" is stronger in the south, and yet it produced little territory there. Perhaps some other force accounted for Purgatory. (The idea that land mainly exists in the northern hemisphere is not Dante's private misconception: it was the accepted Aristotelian view of his day.)

Pondering all this, Mary closed her notebook, which she'd filled with sketches of the moon and observations about the nature and color of the evening sky. The moon vanished beneath

17. Northern slope of the Mountain of Purgatory, up which Dante climbed. Being in the southern hemisphere, this was the sunny side, and he followed the sun's course, from east through north to west

Fig. 1. Illustration of Dante's Purgatory, with caption, from Mary's book *Dante and the Early Astronomers*.

the Palani Hills. Mary stood and brushed grass from her ankle-length skirt. As she would later write of this day, Dante's error disconcerted her. Still, the joy she felt at living on a mountain-top near a well-equipped observatory, with time at the end of each evening to read and think about poetry, overwhelmed her. Her only irritation was the lack of books out here—studies of medieval history that might help her determine how much of existence Dante got right. Within a few months, she would begin to compose one of the finest books ever written on Dante's science, and a most unusual examination of poetic imagination.

2

To the Lighthouse

The Palani Hills rise above waterfall-filled forests and lakes in southern India where plums and plantains grow alongside rare kurinji flowers that bloom once every twelve years. Kodaikanal, the name of the region, is a variation of a Tamil word the etymology of which is difficult to trace.[1] Thoughts about its precise meaning vary from scholar to scholar, from resident to resident, and range from "gift of the forest" to "shade in the summer" to "summer, a mirage" (in the early Christian era, the word appears in the *Pattupāttu*, a group of ancient Tamil poems, to denote "forests green even in summer").[2] For Mary Evershed, the landscape's mirroring of Dante's geography, which helped her visualize the Pilgrim's trek as she reread the *Comedy*, was indeed a gift. The Kukkal cave mouths, rough, overhanging outcrops in a pine and wattle forest flush with geraniums, wild orchids, and leeches the size of a man's back, formed the Inferno's maw. Gentle, glittering ponds on evergreen slopes recalled *Purgatorio*'s opening lines, in which the poet sails the "small bark" of his wit on smooth waters beneath a sapphire-colored sky.[3] And when Mary hiked past stone temples humming with prayerful chants, reached the hills' highest peak, jutting through rain-fed streams, and came to the Kodaikanal solar research center, she felt she had glimpsed the Empyrean.

Her unlikely presence in this spot was the result of a pilgrimage that, while strictly earthbound, was almost as miraculous as Dante's own. Born on the first day of the year 1867 in Plymouth Hoe, England, she was educated entirely at home. Her father,

Andrew Orr, an artillery officer, didn't value formal schooling; that said, he also didn't think girls should be treated any differently than boys. Mary and her little sister, Lucy, were given as much access as their brothers to the harsh governesses who dispensed the family lessons. (There were seven children in all, including two other older sisters.)

Mary's father died when she was three. Soon afterward, Mary's mother, Lucy, took her sons and daughters to live with her father, a clergyman in a country vicarage near Bath. Although the family had suffered a personal tragedy, Mary's father had left them comfortable financially, and he had instilled enormous pride in the children. Mary would not be the only Orr to secure an impressive place in the world. Her brother Charles would become the governor of the Bahamas in 1927.

A young governess, Miss Hawar, assumed responsibility for the children's mental development, insisting that they always look for the best in people. Mary, whom the family called Mindie, loved Miss Hawar and for the next nine years lapped up whatever the young woman said. This was the full extent of Mary's schooling. After that, she was on her own.

Despite no longer living in Plymouth Hoe, she never stopped dreaming of the stormy coastline or the stories her father had told her of the Wreckers, smugglers and plunderers living among the rocks on shore who, with lanterns and mirrors at night, mimicked the pulsing of a lighthouse to lure cargo ships to their doom in the wave-bucked Sound, then made off with casks of spirits and reams of damp tobacco.

An actual former lighthouse, Smeaton's Tower, was reerected overlooking the Sound the year Mary turned fifteen. It became one of her favorite landmarks whenever she'd return for a visit to the Hoe. Made of lime and modeled on the shape of an oak tree, broad at the bottom, narrow at the top, the tower had first been built in 1759. It stood originally on the Eddystone Reef,

Fig. 2. Smeaton's Tower, Plymouth Hoe. Courtesy Shelly Chapman, Getty Images.

south of Plymouth Sound, seventy-two feet above the sea. Initially, it contained twenty-four candles. A timepiece installed next to the light chimed every half hour, signaling the lighthouse keeper it was time to replace the guttering flames.

In 1877, workers discovered that the granite ledge beneath the lighthouse was swiftly eroding. The structure shook every time a wave hit the reef. The tower was dismantled in 1882 and moved to Plymouth Hoe.

By that time, drawn by natural beauty and encouraged by Miss Hawar, Mary had become enamored of the night sky and of the instruments—mirrors, lenses, clocks—employed to observe it. Naturally, the old lighthouse, spiritual sister of an observatory, enchanted her. She would sit against its sloping base and stare across the Sound. There, her father had told her, the Pilgrim Fathers set out for the New World aboard the Mayflower, not knowing what they would discover.

Memories of her father and of the glorious days in the salt air that she had shared with him, staring out to sea, listening to stories of adventure, instilled in Mary a longing for travel (a sense of unfulfilled yearning always associated with her father) and a love of the outdoors unusual among young girls she knew. Girls of Mary's station were raised to be socially skilled by learning ballroom dancing and how to set proper tables in cozy oak dining rooms. Having been schooled at home, she had missed formal education in music, elocution, and deportment. Under her mother's tutelage, she observed social etiquette and wore the required pleated satin dresses but showed little patience with the "bearing and carriage" considered "features of the Gentlewoman."[4] Although she inherited from her mother a fondness for domestic fineries, she much preferred sitting outside in the evenings, in the shadow of Smeaton's Tower.

Her first lighthouse, the splendid Godrevy, graced the Cornwall coast at St. Ives, a small fishing village a few miles away. It was built just nine years before Mary was born, and her father had taken her to see it as a girl. It stood 980 feet above the sea, a white octagonal tower, its clock-driven light flashing every ten seconds, visible fifteen miles distant across the ocean.

When Mary was five, Virginia Stephenson's father began renting the Talland House in St. Ives every summer for his family. Like Mary, Virginia loved spending time outdoors, sunning on the sand or hiking along steep, grassy cliffs. Mary never met Virginia, as far as we know, but they both signed the guest register at the Godrevy lighthouse, and as girls they would have shared a similar vision of the world. Mary loved to read. Virginia's father was a scholar in the natural sciences; from him, Virginia developed an interest in astronomy, which led her to acquire a small telescope as well as express an abiding curiosity in cosmological theory. In 1927 she would use the Godrevy as the model for the central image in her novel *To the Lighthouse*,[5] an achievement roughly contemporaneous with and as radical in its use of time as Einstein's theories of relativity, the problems of which would absorb Mary and her husband in India.

3

The City of Stars

Victorian academics had long extolled the virtues of Italian literature. In the decades following 1818, after English translations of *The Divine Comedy* began circulating widely in the United Kingdom, Dante became well known to British schoolchildren. When Mary was twenty, she and her sister Lucy traveled to Florence to extend their educations. They visited the Baptistry where Dante had laid his head against the wall. Mary confessed to her sister how terribly moved she was that Dante, in exile, longed to see his beloved Bel San Giovanni just once more, a wish he was never granted. In the Bargello, she stared at the intelligent young face modeled on Dante's in the portrait by Giotto of souls in Paradise. Mary stood as if waiting for Dante to speak to her.

John Ruskin once wrote that, of all poets, Dante was the most sensitive to light's effects on the eye.[1] If Dante could say anything at all to Mary, she wished it could be an explanation of the light, so much more intense in Italy than it was back home. She was as struck by Florence's sky as she was by its buildings, streets, sundials, and bridges. She had heard it called the City Founded on the Stars. She knew the old fable—a fable Dante would have read—which said that after God had sown confusion on Earth following the collapse of the Tower of Babel, Europe was inhabited by the descendants of Japhet and the third son of Noah. One of the most notable of these descendants was Atlas. According to legend, Atlas came to Tuscany with his wife, Electra; searching with the aid of astronomy through all the regions of Europe for the finest place to live, he had chosen to settle on the Mount of Fiesole.

From Florence Dante would claim to rise to the holy stars—an imaginative journey that may have begun when he was a child studying the starlike mosaics inside the Baptistry.[2] In 1225, forty years before Dante was born, a Franciscan friar named Jacobus had begun the *tribuna* mosaic inside the building. In the following years, across the eight-sided cupola, visions of the Last Judgment and the Nine Orders of the Angels would appear. Further mosaics would be added to Bel San Giovanni when Dante was ten—all still extant during Mary's visit to Florence in 1887. In particular, she thought, the *tribuna*'s large round portrait in which the Virgin Mary and John the Baptist sit enthroned in a flowerlike design must have impressed the young poet, as evidenced by its return as one of the strongest images in his greatest work.

A few blocks from the Baptistry, on a side street near Via Calzalli, the Chiesa di Santa Margherita de' Cerchi rose in perfect modesty. The Portinaris, Beatrice's family, worshipped here. In the *Vita Nuova*, Dante wrote of one afternoon when he sat, a lovesick young man, holding a tablet in his lap, sketching from memory his lovely Beatrice as an angel.[3] He may have claimed a spot on a wooden bench right next to the little church. Mary imagined him slouched intently over his drawing, his hand steady, his lines as light and precise as a geometer's grid.

Inside the Chiesa, she was instantly reminded of the elaborate astronomical image of the sun and moon in equilibrium, "making a belt of the horizon," in *Paradiso* XXIX.[4] A chill filled the chapel, and a small window high in the wall above the altar teased Mary's eye—too distant to admit any vista it might promise, provoking a mild spiritual yearning in the viewer. Its small size dimmed the sunshine spinning through it, turning the light milky in the room, more like moonglow. It was as though

Fig. 3. Dante in his city. Courtesy Darragh Hehir, Getty Images.

morning and evening existed simultaneously inside the chapel. In Dante's passage, the sun settled low in the constellation Aries (the Ram), with the moon poised opposite it, in Libra (the Scales): an absolute balance, both timeless and fleeting.

From her grandfather's sermons, Mary remembered St. Augustine. He had referred to the angels' "twilight knowledge," the contemplation of their own natures, as well as their "morning knowledge," their contemplation of God.[5] When the Book of Genesis described Creation, noting, "There was evening and morning on the same day," it was, in fact, defining the angelic consciousness.[6] Thomas Aquinas, in the *Summa Theologica*, a book Dante knew well, insisted that angels understood everything immediately at birth. To illustrate this possibility, Aquinas had used an astronomical image: "changes can happen at the same time and in the same instant, just as in the same instant that the moon is [lighted] by the sun, the air is [lighted] by the moon."[7]

Moving farther inside the church, Mary ran her hand along the cold stone wall where the Portinari family tomb was said to be. Whether or not Beatrice was actually buried here—many scholars think it unlikely—she imagined Dante standing, full of grief, in front of this ancient mausoleum. It was shaped like a box made to cradle a musical instrument. From the small, blue, chalky encasement in the wall, his mind, like an angel's, opened out into infinity, an imagined reunion with his love in the starry vault of Heaven.

In another church Mary visited with her sister, the Basilica of Santa Maria Novella, a Dominican priest named Tommaso Caccini once served. Mary had read that, in 1616, this man testified before the Holy Office that Galileo Galilei was a heretic for supporting the Copernican system.

One afternoon, on the hillside of Bellosguardo, Mary toured the Villa dell'Ombrellino, where Galileo studied sunspots with his telescope. Her education, both in the life and poetry of Dante and in the history of astronomy, was coming into focus.

Before leaving Italy, she and Lucy visited Dante's tomb in Ravenna: "Here I lie buried, Dante, an exile from my birthplace, a son of Florence, that loveless mother. . . My soul was taken to a better place and dwells in bliss with its creator among the stars."

4

Poetry and Sunspots

In Ravenna, Mary truly began her lifelong conversation with Dante, and established the intellectual foundations that guided the rest of her life.

Here is some of what she learned.

Sometime in 1318, at the age of fifty-three, Dante arrived in Ravenna, apparently at the invitation of the city's ruler, Guido Novello da Polenta. For almost twenty years, Dante had been wandering in exile through Tuscany, Romagna, and the Veneto. Nasty politics in his home city, fueled by refusals to compromise and by lies launched by political factions against their opponents, had fostered an atmosphere of civil war in Florence. Caught in the civic crossfire, Dante was banished from his home and threatened with execution. His assets were seized, his property condemned.

His response to this injustice was to create *The Divine Comedy*, a vessel for expressing his angers and frustrations as well as his hopes and aspirations. It was not the first poem he had written. Since earliest youth, he had studied and practiced the lyric poetry that flourished in Provence beginning with the work of William of Poitou (1071–1127), in which "there emerges a concept of sexual love previously unknown in literature," the literary scholar Thomas G. Bergin writes. "The poet's attitude toward his lady is characterized by reverence, timidity, and awe, with occasional flashes of hope and successful courtship. The lady is almost unattainable, and the poet's worship of her is a source of torment and melancholy pride."[1] By the time he was eighteen, Dante considered himself a skilled practitioner of Provençal poetry. In his greatest works, the *Vita Nuova* and *The Divine Comedy*, he would transcend the Provençal tradition by fully equating earthly and spiritual love (his unattainable lady, Beatrice, becomes his soul's savior and his guide to the afterlife) and by including politics and natural philosophy in his work.

The Divine Comedy insisted that he, Dante Alighieri, a middle-aged man stumbling, lost, across the countryside, somehow journeyed through Hell, Purgatory, and Heaven, then returned to tell the tale so that his fellow citizens, trapped in selfish pursuits and deadly politics, could benefit from what he had seen and secure God's blessings. That he managed to compose the *Comedy* while on the run from his enemies and in search of generous patrons, always near libraries where he could do his research, astonished Mary. Just as he cast his antagonists into Hell in his verses, he flattered powerful people who could help him, leaders such as Cangrande della Scalla of Verona. In *Paradiso* XVII, Cangrande is described as a munificent man through whom "the fate of many men shall change, / rich men and beggars changing their estate."[2] On a certain level, Mary saw, Dante's poem was a sly, pragmatic document, offering social commentary and political cover for his work.

His flattery succeeded only intermittently. Like most politicians, Cangrande did not recognize the utility of artists. Dante was welcomed, when he was welcomed, because he was a literate man, capable of writing official decrees. He had gained some limited political experience in Florence, so rulers could dispatch him usefully on diplomatic missions. As for his poem—well, perhaps it was Cangrande's discovery that in *Purgatorio* Dante had called his father a tyrant that eventually got Dante booted from Verona.

By the time he came to Ravenna, handwritten copies of *Inferno* and at least parts of *Purgatorio* had circulated throughout the countryside. For better or worse, he had become famous. In Ravenna, a Franciscan friar associated with the Office of the Inquisition confronted the poet in church. "Are you the Dante who says that he went to Hell and Purgatory and Heaven?" he asked. Dante replied, simply, "I am Dante Alighieri of Florence."[3] The friar warned him to stop writing blasphemy (in the vulgar vernacular, no less) and to compose in proper Latin a paean to God's glories.

Rumors that Dante had visited the Underworld persuaded even some educated men such as the Lord of Milan that Dante practiced black sorcery. Matteo Visconti, the Milanese ruler, opposed Pope John XXII. He considered enlisting Dante's powers of necromancy to murder the pope.

Dante did not discourage wild rumors. It pleased him to overhear women in the streets of Verona or Ravenna whisper that he came and went at will from Hell's scabrous crags, returning to Earth with news of the dead. You could tell the story was true, some said—just look at the poor man's singed beard, his brown complexion.

His Ravenna patron, Guido Novello, was a rare man, Mary learned—a poet, a lover of literature, as well as a political despot. Boccaccio, Dante's first biographer, writes that Guido Novello had "for a long time known" of Dante's "worth by reputation," and he valued the poet's presence in his city precisely because of his intellectual merits.[4] Unlike Cangrande, Guido Novello saw that embracing an artist respected for eloquence, discipline, and principles might enhance his own name. Furthermore, if Ravenna became known for supporting cultural work and developed into an artistic center, its economy might grow. (According to Boccaccio, Dante did, in fact, attract people to the city, from as far away as the university in Bologna, who wanted to study rhetoric with the "divine old man.")[5] Guido Novello was shrewd enough to consider art a resource. He had developed libraries in the city, notably the Cartilegio, with a collection of historical records as fine as any on the eastern half of the peninsula; as a result, for the first time since his expulsion from Florence nearly two decades earlier, Dante enjoyed serenity and unpressured time to finish his vision of the universe.

From Mary's time to ours, we have valued *The Divine Comedy* not least because it catalogs the sum of human knowledge in Dante's day. It gives us glimpses of what we have built on and what we still must embrace (or escape) to advance ourselves in science, art, theology, politics. The poem brings us news of love and war as they were practiced in the past. Guido Novello's libraries, his subsidies, his gift of time to Dante, were major contributions to our history and our self-awareness.

As did Mary, today's visitor to Ravenna will hear, each dusk, the bells of the basilica of San Francesco—the church to which Dante's tomb is attached—chime in honor of the opening lines of *Purgatorio* VIII, one of the most beautiful passages in the *Comedy*:

> It was the hour when a sailor's thoughts,
> the first day out, turn homeward, and his heart
> yearns for the loved ones he has left behind,
>
> the hour when the novice pilgrim aches
> with love: the far-off tolling of a bell
> now seems to him to mourn the dying day—[6]

It is not the dying of the day but its predawn that is likely to strike a traveler to Ravenna. In early spring, the city's birds serenade the sky's cracking light at four A.M., cooing and trilling, whistling in tercets—swallows, larks, magpies, mourning doves. "A bird quiet among the leaves she loves / sits on the nest of her beloved young, . . . / foretelling daybreak from an open bough, / she waits there for the sun with glowing love," Dante wrote in

Paradiso XXIII.[7] He called birds the little brothers of Saint Francis. Their chattering formed the lulling rhythms he listened to as he imagined Paradise. Early in the morning, wingbeats hidden in shaded trees might convince a weary immigrant of the presence of angels, as bells amplified other bells' echoes throughout the city.

Mary understood that Dante the poet possessed the impulses of a scientific researcher: a restless mind, a habit of close observation. Studying Ravenna's stunning mosaics, glittering shards of gold and glass and stone gilding the interiors of the city's cathedrals and mausoleums, he could determine which species of birds flitted about the marshlands and forests surrounding Ravenna in the fifth and sixth centuries, when many of the mosaics were made. Rock partridges, ring parrots, doves—their images fill the San Vitale cathedral, the Basilica di Sant'Apollinare Nuovo, the Mausoleum of Galla Placidia. Not only are the birds' winged aspects on the ceilings and walls delightful, but their abiding presence in these sacred spaces inclined Dante to view them as perfect emblems of eternal life. The scientist in him noted their survival through the centuries; the poet viscerally conveyed the *concept* of survival.

Time and again in *Purgatorio* and *Paradiso*, Dante employs bird imagery to evoke Heaven's endless bliss, as in *Paradiso* XX:

> . . . like the lark that soars in spacious skies,
> singing at first, then silent, satisfied,
> rapt by the last sweet notes of its own song,
>
> so seemed the . . .
> reflection of God's pleasure, by Whose will
> all things become that which they truly are.[8]

Ravenna's mosaics, like those in the Florentine Baptistry, offered Dante a rich trove of patterns, colors, and reflections with which to turn the abstraction of Paradise into bright, concrete imagery.

Accordingly, Mary became enamored of mosaic techniques. Talking to laborers in the city's old workshops each day, she discovered that mosaics date to the third millennium B.C., when monochrome pebbles were used to pave the courtyards of Crete's royal palaces. She was told that mosaic-makers would not have labeled themselves scientists—they would have thought of themselves as artisans—but their process of production depended on what modern observers would call scientific study of the Earth and its materials, as well as engineering methods subject to testing, refinement, and successful repetition. Usually, the process began with a sinopah, a preliminary sketch made using clay quarried from the city of Sinop, in modern-day Turkey. Fresh plaster would then be applied over the sinopah, after which artisans would press, by hand, single pebbles, chips of gold, and cut glass made of sand into the mortar, assuring a beautiful, uneven symmetry generating multiple reflections.

Intrigued, Mary studied the practices of imperial courts, how the courts gathered scholars from all fields and all countries within their cloistered walls: the mosaics' themes were intended to encompass the entirety of human knowledge—art informed by the sciences of astronomy, medicine, theology.

In constructing *his* universe during his last days in Ravenna, surrounded by dazzling mosaics, Dante followed a similar all-inclusive blueprint. "Look up now, Reader, with me to the spheres; / look straight to that point of the lofty wheels / where the one motion and the other cross, / and there begin to revel in the work / of that great Artist who so loves His art, / His gaze is fixed on it perpetually" (*Paradiso* X):[9] this heavenly ecstasy could well have been inspired by the vault of the Mausoleum of Galla Placidia, with its rows of green and yellow circles, glowing starlike, connected by tresses of blossoms, as if light had become

Fig. 4. Mosaics in the vault of the Mausoleum of Galla Placidia, Ravenna. Photo by Tracy Daugherty.

a garden bursting into bloom. Mary loved the mausoleum. Its inlaid stars were surrounded by birds and angels and fountains and ribbons, by burning sunlight piercing the amber alabaster windows, an overflow of every phenomenon human senses could tolerate.

In the fourteenth century, as in Mary's day and now in our own, nasty politics were cancerous, nearly always fatal. They undermined the progress an individual community could generate. Even in Guido Novello's Ravenna, a place that revered the arts and sciences, Dante was finally called upon to prove he was useful, politically. His patron sent him on a diplomatic mission to Venice, and it killed him.

The Doge of Venice had accused Ravenna of attacking Venetian ships. Dante was sent to the watery city to broker peace, a mission that failed, though eventually the two municipalities did avert war. This was not Dante's first trip to Venice. Mary knew this from her reading. Early in his exile from Florence he had traveled there and—with his eye for precision—marveled at the Arsenale, the city's intricate shipbuilding complex, an engineering miracle using canals to create the world's first assembly line, mass-producing Venice's vessels of war and trade. In another instance of employing details from the world of science to enrich his art, Dante evoked the Arsenale to explain the nature of the barrators' punishment in Hell:

> In the vast and busy shipyard of the Venetians
> there boils all winter long a tough, thick pitch
> that is used to caulk the ribs of unsound ships . . .
>
> here, too, but heated by God's art, not fire,
> a sticky tar was boiling in the ditch
> that smeared the banks with viscous residue.
>
> (*Inferno* XXI)[10]

Into this vile pitch the barrators (the sellers of Church offices) were plunged.

Visions of punishment must have plagued Dante while he pursued his diplomatic mission. He was not feeling well. He was exhausted, having just completed *Paradiso* in a fever of concentrated productivity. He did not welcome his return to politics, but he could not insult his patron's generosity. As he walked, shoulders hunched, through Piazza San Marco between the Duomo and the towering campanile (a former lighthouse), with its carvings of angels and its ancient iron bells; as he sat in the dim waiting room of the Doge's palace, anticipating his meetings with representatives from the Council and the Senate, he must have felt satisfied, in spite of his growing discomforts, that he'd gotten the world exactly right in his *Comedy*. As did he, Mary could still see in the plaza, in the palace, in 1887, the feast of human folly vividly displayed: the vanity and greed of Church and State. Aside from its grand chambers, lavishly adorned with paintings and jeweled furnishings, the Doge's palace contained cold, dark prison cells where moaning heretics, shackled to the walls, perished in agony—in a single building, humanity's tragic history, stark manifestations of Heaven and Hell.

After failing to engage Venice's leaders in useful dialogue, Dante returned to Ravenna in a flat-bottomed boat through the Po delta's turgid lagoons. The pale luster of the Adriatic in the near distance undulated dully in the moonlight. It was the deadly season: early fall. An anopheles mosquito bit him. Soon, malaria felled him. In Ravenna, on the night of September 13 and the early morning of September 14, 1321, while swallows piped in the eaves outside his little room, Dante died.

Guido Novello arranged a splendid funeral for him in the basilica of San Francesco, where the poet's body was first interred. (The ruler's plans to construct a magnificent tomb for Dante fell through when nasty politics swept away his powers and he was forced to flee the city, shortly after Dante's death.) Today, visitors to the basilica can descend a short set of steps to peer into the large, dark crypt where Dante was buried, with its faded mosaic floor, its cream-colored columns. For centuries, the church has been flood-prone; now, Dante's old resting place is a goldfish pond. It seems the local Franciscans are not immune to scientific methods: years ago, the priests introduced fish into the crypt's placid green pool to serve as indicators of the water's toxicity levels.

In 1780, Dante finally got his promised tomb in Ravenna, when the cardinal legate Luigi Valenti Gonzaga commissioned a small but ornate mausoleum to host the bones of the man even Florentines now regarded as Western Europe's greatest poet. On a sunny summer morning, Mary stood reading and rereading the ceremonial poem written by Dante's friend and admirer, Giovanni del Virgilio of Bologna. It was etched into stone inside the shaded tomb, and it celebrated the poet's glorious "return to the stars."

Two hundred and eighty-eight years later, with Dante on his mind, Galileo Galilei turned his telescope to those very same stars. He did not invent the 'scope. A German-Dutch eyeglass maker, Hans Lippershey, is said to have noticed two children playing with lenses in his shop one day in 1608, holding two glass pieces together to better see a weathervane atop a nearby church. From watching the kids, Lippershey devised a spyglass.[11] When Galileo heard of this invention, he determined to improve upon it, fitting refracting lenses into a metal tube and placing this contraption on a steady stand. Galileo ground the glass using artillery balls and ordered organ pipes to be used as the initial bodies of his telescopes.[12]

In her travels to Florence and Venice, Mary was delighted to discover that Dante gave Galileo his first big career boost. In 1588, the young scientist delivered two lectures to the Florentine Academy. To bolster his scientific bona fides, he relied on the authority of art. The lectures were on the "shape, location, and size of Dante's *Inferno*."[13] Galileo had taught himself the mathematics of Euclid and Archimedes. When the chair of mathematics at Pisa opened up, Galileo was invited to speak at the Florentine Academy. Essentially, this was a job interview (Galileo was twenty-four years old at the time, a medical school dropout).

The Academy had been founded by the Medici dynasty. Its purpose was to glorify the Medicis in every intellectual arena;

like Guido Novello, the Medicis understood the arts and sciences as resources, useful political tools. Although they exploited lavish cultural displays to increase their prestige, they did so with a genuine love of the arts and with the culture's overall health in mind. In particular, the Academy was concerned with the Italian language—which grammatical elements should claim its core? (This question had obsessed Dante as well.) Ostensibly, then, Galileo's lectures concerned literature and language, though he took as his topic the challenge of mathematically rendering the dimensions of Hell.

Many of his listeners considered it improper to apply scientific methods to a work of art, just as the Church's political leaders would later insist that Galileo not meddle, scientifically, with theology. It is a testament to Dante's status as cultural icon that Galileo knew he needed the poet if he was to be taken seriously as a scholar.

In his lectures, Galileo endeavored to sketch Hell's architecture, based on Dante's descriptions, geometrical theorems, and the latest measurements of Earth. Numerical proofs, he said, indicated it was possible for Dante's imaginative construct to support itself, much the way Brunelleschi's dome stood firm on the Florentine Duomo. Galileo didn't quibble, at the time, with Dante's depictions of giants or of Lucifer, massive, encased in ice at the center of the Earth. By taking Dante seriously, Galileo gained insights into important principles of structural engineering, which would fascinate him further six years later when, like Dante, he spent weeks studying the Venice Arsenale.

In 1593, he was hired as a consultant at the Arsenale, advising military engineers and instrument makers on ballistics and propulsion. By then, he had read Aristotle's *Mechanical Questions* and wanted to test the philosopher's theories against real-world applications. His experience around the shipbuilders taught him that he had made a mistake in his Dante lectures. He understood now that objects could not be scaled up, simply. A man cannot become a giant just by making each of his appendages larger. Structural supports, such as the individual bones in a human skeleton, must grow proportionally thicker as limbs get bigger, otherwise the construction will collapse of its own weight. A wooden ship pulled from the water risks breaking apart. Nature, Galileo concluded, is not scale-invariant. The material involved is crucial to scaling.

This theoretical breakthrough was another boost to his career. By 1609 he had developed an eight-power refracting telescope, and he was ready to demonstrate it for the Doge of Venice and members of the Venetian Senate, in what amounted to another

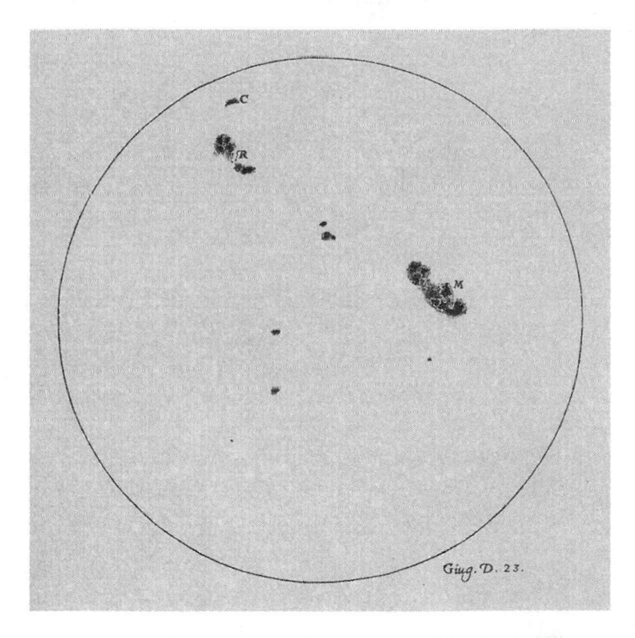

Fig. 5. Sunspot drawing by Galileo Galilei from *History and Demonstrations Concerning Sunspots and Their Properties* (1613). Courtesy Galileo Project.

job interview. Together, the men climbed the slender, circular iron staircase inside the campanile at San Marco (Galileo might have noted how eerily similar to the ascending terraces of Dante's Purgatory this frightful staircase was), and then he pointed his 'scope out to sea. The Doge gasped as distant ships came rushing into his eye. It was as incredible as if he'd witnessed Charon ferrying dead souls across the River Styx.

For the politicians, scheming their nasty business, the telescope was mostly a weapon of war, but with it Galileo would upend the universe. His discovery of sunspots, using the instrument, would shatter the notion of heavenly purity, so central to the Aristotelian-Ptolemaic cosmos that Dante had codified in the *Comedy*. Humankind's understanding of its place in the universe, and therefore its self-awareness, would take another step forward.

In rewriting existence, Galileo resorted—as he had at the beginning of his career—to Dante's rhythms, rhetoric, and imagery. His descriptions of the sun in *Letters on the Sunspots* (1612) echo Beatrice's discussions of the moon's black patches in *Paradiso* II. Dante's insistence that the moon had taken him into its substance "as water takes in light, / its indivisibility intact,"[14] is absorbed into Galileo's argument that lunar dark spots could not be caused by chasms *inside* the moon: even "the clearest waters are not so utterly limpid and transparent that they would allow our eyes [to reach] such depths," he wrote.[15]

Mary caught the echo of Dante in Galileo. She understood from her observations and her reading that a scientist builds upon the work of his forebears. Additionally, she saw, the scientist and the poet share a problem: that is, finding a language to represent what lies beyond representation (such a vocabulary often depends on contrast, light and dark, for which the celestial realm is tailor-made). Mary grasped that, throughout his evolution as a researcher, Galileo always considered Dante a peer, his artistry a model. He did not see himself disputing or erasing Dante's achievements as a chronicler of cosmic possibilities; instead, he was adding to the picture, shading in details. For Galileo, *The Divine Comedy* was not a graveyard of discarded astronomical theories, but a poetic prologue to future discoveries. It took both the artist and the scientist to bring *what is* into focus.

As she packed to leave Venice, Mary was struck by an astonishing thought: no one now remembered the names of the small leaders who persecuted Dante and Galileo. Many of the political boundaries, the strategic alignments that existed in 1321 and 1609, had long been obsolete.

Yet essentially, Mary knew, we were all still living in the universe Dante and Galileo laid out for us; if our understandings of structures, dimensions, and causes had changed over time, our sense of wonder at the vastness and patterning of existence, spurring us to new discoveries, remained as rapturous as the expressions of the poet and the scientist in *The Divine Comedy* and Galileo's *Starry Messenger*.

In 1519, Florence's political leaders, retrospectively ashamed of their treatment of Dante and dismissal of his art, lobbied to get his bones back. By now, most of the bones had crumbled to dust, but in the world's imagination they had thickened considerably. Dante had become a giant.

Mary read that Ravenna workmen, consulting scientists and engineers, had devised a plan to hide the poet's remains in a secret chamber behind a hollow wall in his tomb. That way, Florence's agents couldn't steal them. In his life a homeless immigrant, Dante now enjoyed his final rest in the place where lively birds alerted him to the possibilities of Paradise.

5

"Black Star-Lore"

Following their travels in Italy, Mary and Lucy spent several weeks in Germany to further their educations abroad. At the time, late 1887, Mary had not yet heard of Albert Einstein, but in Berlin, in the southern part of the city known as Schöneberg, she walked narrow streets destined to spawn rows of identical concrete apartment buildings. One day, one of these buildings would house the man who would shatter former conceptions of the universe. He would disprove Isaac Newton's notion that "there must be a drawing power in matter."[1]

Famously—perhaps apocryphally—Newton was sitting in his mother's Lincolnshire garden one afternoon in 1666 (forced to retreat there by the plague, raging in public squares). Mary knew the story well and so did Einstein. Newton was contemplating the moon's orbit around the Earth when an apple was said to have fallen from a tree onto his head. From this incident, Newton developed the idea of gravity as a force generated by mass, akin to invisible grappling hooks. The gravitational attraction between two bodies, he said, was proportional to the product of their masses.

At the time of Mary's visit to Berlin in 1887, Einstein's challenge to Newton's ideas still lay many years ahead, but the city, noisy and bustling, so much more vibrant than sleepy Cornwall and the ancient Italian towns she had just explored, seemed to her to belong to the future. She was eager to move forward. A rare photograph of Mary, taken when she was just back to England, shows a slender woman with a long, graceful neck

Fig. 6. Mary Evershed, age 20. Courtesy Science Museum Library / Science & Society Picture Library.

trimmed with a ring of tiny pearls, staring keenly at a weak source of light. She looks bemused, slightly impatient, her brown, curly hair piled thickly above a small, round face. Her lips are full, her eyes dark. She poses stiffly, as though she can't wait to shuck this heavy satin dress with its puffed sleeves, a relic of a bygone era (nothing like the sleek, airy dresses she had seen in Berlin), and hurry back out to chart the sky.

Three years later, the sky again captivated her when she, young Lucy, and their mother joined an older sister for a five-year stay in Australia, where the sister had settled in marriage—at precisely the spot, Mary knew (32 degrees south), where Dante had located Purgatory as an island in the center of the ocean in the antipodes of Jerusalem. They were in New South Wales, where the British government had established a penal colony in 1787. By 1850, with the discovery of gold and rich farmland, the region had become a favorite destination for immigrants feeling dissatisfied with life in the British Isles or looking for adventure. "How beautiful is the whole region . . . the hills, the capes, the promontories; and then, the splendor of the sunlight," Mark Twain wrote after briefly visiting the place. "And it was in this paradise that the yellow-liveried convicts were landed . . . a sort of bringing of heaven and hell together."[2]

Near the house where Mary settled with young Lucy and their mother, Australia's best-known astronomer, John Tebbutt, ran a private observatory in Windsor. Tebbutt had discovered the Great Comets of 1861 and 1881. Mary made regular visits to his observatory—no small feat, given his distaste for guests. His impatience with people in general had steadily worsened and had reached a bursting point by the time Mary introduced herself to him. His journal entries, in the years prior to her arrival, trace an irritability Mary would encounter in *many* male astronomers: "I did not observe the comet last night owing to the Reverend John Mosley being at the observatory" (1864);

"Last evening a Windsor gentleman came to the observatory according to appointment to see the moon, but instead of bringing only another visitor with him as stipulated he brought three others. . . . The consequence was that my small round equatorial chamber was well-filled. The accident which happened to me on the 8th [the 'spider hairs on the micrometer eyepiece' broke] was due to the agitation produced by a visitor at unseasonable hours. . . . I endeavor to avoid visitors as much as possible [now], for I find there is nothing but ill luck in the reception of them. They come without five minutes' preparation for what they wish to see, and go away just as edified as before their arrival" (1891).[3]

Mary instantly distinguished herself from his other guests, impressing him with her insightful questions. He saw she was a serious young woman, apparently uninterested in what he considered frivolous social affairs, the many dances and sports events held in the area each week throughout the year. Though without the benefit of any formal schooling, she was already astonishingly knowledgeable about astronomy. Tebbutt encouraged her to extend her studies and gave her remarkable access to his observatory.

As she had in Italy, Mary discovered an appetite for the history of astronomy as well as its practice. In New South Wales she met Katie Langloh Parker, an Australian woman of English descent. Parker had collected oral folk tales of the Gamilaraay Aboriginal people near the whaling fisheries at Encounter Bay, South Australia, and on the sheep farms along the Darling River in New South Wales. She wrote the stories up and later published them for British audiences. From Parker, Mary learned that Australia's indigenous peoples were probably the first human observers to catalog the constellations and give them names.

Mary loved the "quaint tales" of the "black-fellows," as she called them, and, back in England, prepared a detailed talk for one of the earliest meetings ever held of the British Astronomical Association, convened at Sion College in London on November 30, 1898.[4] The meeting's secretary recorded her name in the minutes as "Miss M. A. Orr." Then thirty-one years old, she was the gathering's sole female presenter, though she was not allowed to read her own paper (even among British amateurs, astronomy was very much a boys' club). Instead the piece, "Black Star-Lore," was recited by a "Mr. Schooling." When he had finished, the BAA members, including Walter Maunder, the organization's founder, having sat all evening through dry and often dreary discourses on Mars, meteors, comets, and the zodiacal lights, praised Miss Orr for introducing them to "these early astronomical legends": the stories were "the most interesting things [the audience] could have [received]," Mr. Schooling enthused. Maunder remarked that he was "greatly indebted to Miss Orr" for her history lesson; knowledge of astronomy's evolution was invaluable in grasping the nature of the science.[5]

Mary's piece reflects her naïve, unconsidered European perspective. Blithely unaware of the depth of her bias, she says the "black-fellows'" narratives, passed on "to their piccaninnies ... prove that this low race has much imagination and some power of observation." The tales' history "throw[s] some light on the fascinating problem of the birth of astronomy and the origin of the constellations."[6]

From Katie Parker she had learned that the Aborigines regarded the sky "as a solid vault, not very high above their heads, for in these tales things thrown up at it stick there, or break, people are described as climbing up into it, or running about over it."[7]

The natives had fastened names to Venus (they called it the Laughing Star for its playfulness, appearing now in the morning, now at night), Jupiter, and Mars, but they did not seem to recognize Saturn or Mercury, Mary observes. They called the moon Bahloo—he was the hero of many legends. Meteors were the spirits of wizards. What appeared to European Christians as the Southern Cross was to the Aborigines a featherless emu devil.

In Australian mythology, the sun was a woman in constant pursuit of the moon. The moon mocked the sun and accused her of entertaining too many lovers. On those occasions when Bahloo overtook the sun, day went dark: it is "remarkable," Mary notes, that the Aborigines had "discovered that the moon is the cause of solar eclipses."[8]

In 1866, the British science publisher James Gall had printed a pocket-sized book with simple maps of the northern constellations. Mary had used Gall's book to learn her way around the northern sky. No similar volume existed for the southern hemisphere. Mary began to document her observations and make charts of Australia's stars, modeled on Gall's maps. Her purpose, she told Tebbutt, was to get "people (children and adults) on the track of observing for themselves the movements of the heavenly bodies."[9]

In 1895, Gall & Inglis, by then a well-known, if specialized, publishing firm, agreed to print Mary's work and produced gorgeous, deep blue maps of the stars, all in a circular shape as if seen through a telescope. Mary called the book *An Easy Guide to Southern Stars* and published it under the gender-neutral name M. A. Orr.

She convinced Tebbutt to provide a preface, hoping his blessing would get the book noticed. Though he states his "pleasure in commending this little work," his inevitable bias against Mary as an amateur and as a woman is apparent in the preface's brevity and thrust. "The whole of my spare time being absorbed in astronomical work of a technical character," Tebbutt writes, "I have not had the opportunity for going into a detailed examination of the contents, but I believe we may safely trust the accuracy of the enterprising authoress. The work is pleasantly written."[10]

The book sold moderately well. Mary's enthusiasm for the development of astronomy, the lore of star names, and the joy of observing the constellations' "most graceful shapes" proved irresistible to those who came across the volume. She offered solid historical background: we "find Orion and the Pleiades in . . . Homer, and the Scorpion and Twins on old clay tablets from Ninevah," she wrote. "These ancient astronomers evidently lived in the northern hemisphere for they left unnamed all stars further south than the ship Argo." She convinced northern readers to shift their mental perspectives: "in these [southern] latitudes Orion seems to be diving downwards, not standing upright . . . and the three little stars which should, as his sword, seem to dangle from his belt are here seen rising up out of it."[11]

By now, Mary was reading Dante in Italian. She carried the *Comedy* with her wherever she went. She told her sisters that in the vast dust-plains of the Australian outback she got a sense of Dante's exile, the arduousness of wandering from city to city. Her initial unfamiliarity with the southern sky helped her see the sky anew, as though she were witnessing stars for the very first time. These were the constellations Dante the Pilgrim saw, emerging from Purgatory "pure" and "remade" and "prepared to rise up to the stars."[12]

As early as 1896, Mary started wrestling with astral problems in Dante. In a tiny notebook she sketched the positions of the sun, moon, and stars as Dante described them. Of line sixty-four in the fourth canto of *Purgatorio*, "il Zodïaco rubecchio," she wonders if Dante means "the zodiac" has been "reddened by the sun" or if he is pointing out a red-hued patch of the "zodiacal cogwheel."[13] Elsewhere, she worries that, while mapping Paradise, Dante has forgotten the equinox. In *Paradiso*'s third canto,

Dante calls the sphere of the moon the "slowest" in the heavens. Mary can't determine if this is a reference to the moon's size or to "diurnal motion."[14] In an agitated hand, she returns to *Purgatorio*, where Dante suggests that, from a particular position in the sky, the sun "lights alternatively each hemisphere." It would be more accurate, Mary writes, to say that it "lights the starry heavens above and earth and three planets below."[15] Dante had found his astronomer.

6

Physical Astronomy

In Mary Evershed's lifetime, British women who wished to become professional astronomers encountered insurmountable barriers. To begin with, astronomy was not taught as a stand-alone discipline in British universities; it was a branch of mathematics and physics. At Cambridge, women were allowed to take examinations, but they could not earn degrees until 1923, and could not become full members of the university until 1948. A woman with her head in the stars could only hope for a menial position as a "computer" in an observatory—often some wretched colonial outpost in the middle of nowhere—compiling tedious numbers as parts of research projects for which she would never be credited.

Women fared no better in elite U.S. universities. Typical of this period was Harvard College Observatory's policy of paying its "computers" twenty-five cents an hour, when it paid them at all. Because Professor Edward Pickering, the observatory's director from 1877 until his death in 1919, received no funding from the college for staff and supplies, he depended on amateurs and volunteers. "Many ladies are interested in astronomy and own telescopes," he wrote, "but with two or three noteworthy exceptions their contributions to the science have been almost nothing. Many of them have the time and inclination for such work, and . . . as the work may be done at home, even from an open window, provided the room has the temperature of the outer air, there seems to be no reason why they should not thus make advantageous use of their skill."[1] This passed as a progressive attitude. Pickering believed his use of women as computers benefited the profession while protecting women's fragile health. He would never think of exposing them to the fatigue of nightly telescope observations or cold winter conditions.

On more than one occasion, Harvard hired partially deaf women, figuring they were free of distractions from their tasks. Regularly, the college received shipments of heavy photographic plates taken in the Peruvian Andes, from its southern observatory, Boyden Station, at Arequipa. The plates were hauled by mules across rickety suspension bridges, and transported by boat and train. In Harvard Square, the glass plates were mounted on wooden viewing frames and, later, turned over to the computers to calculate the relative brightness of the tiny star images.

In spite of demeaning pay scales and often brutal working hours, computers made significant contributions to astronomy. In 1882, Pickering reported to Harvard president Charles Eliot that female volunteers had made an astounding "900 measures" of a particular variable star, tracking its magnitude shifts "in a single night, extending without intermission from 7 o'clock in the evening until the variable had attained its full brightness, at half past 2 in the morning."[2] The women's dedication and efficiency in poring over photographs was so great that "for many purposes the photographs take the place of the stars themselves, and discoveries are verified and errors corrected by daylight with a magnifying-glass instead of at night with a telescope," Pickering said.[3] One of his computers, Williamina Fleming, became so adept at reading stellar spectra (the rainbow colors that Isaac Newton had insisted belonged to light itself and not to the corrupting qualities of glass lenses, as many of his peers suspected) that she quadrupled the number of then-known star categories, paving the way for greater understanding of the differences among observable stars, based on temperature, chemical composition, and stage of development.

Internationally, women had no choice but to be amateurs, in the tradition of Caroline Herschel, sister of William, who discovered the planet Uranus. In the 1780s, when William sacrificed a music career to manufacture telescopes, Caroline became his assistant, polishing lenses and jotting down night-sky observations. Together, sister and brother fashioned telescope-molds from wagonloads of horse dung, conjuring heaven from the wastes of the Earth. On her own, Caroline discovered eight comets. After William's death in 1822, she compiled a catalog of nebulae which William's son, John, used to advance his career as a scientist, though he never gave Caroline credit.

In the early nineteenth century, children—and especially girls—became enamored of a series of constellation cards peddled under the name "Urania's Mirror" in England. The cards showed the major constellations with punched-out holes in place of the stars. When held against a light, the holes (of various sizes, to indicate relative stellar magnitudes) could dazzle a child's eyes. "Urania's Mirror" had been devised by an unidentified "lady"—she is referenced in a popular book of the day, Jehoshaphat Aspin's *A Familiar Treatise on Astronomy* (1825)—a savvy amateur whose name may never be known to us.[4]

Among women in pursuit of the sky, sunspots and solar flares were favorite objects of study. Elizabeth Brown, a tireless amateur, suggested that sunspot drawing was a perfect activity for "ladies" because they had plenty of time to devote to it and needed only a dark glass and a pencil and paper to do the work. Perhaps most importantly, through this work their delicate constitutions could avoid exposure to the chilling night air.

(Brown had literary ambitions, publishing books about the sun—anonymously, so they'd be taken seriously—under the slightly lurid titles *In Pursuit of a Shadow* and *Caught in the Tropics*. Generally, science books written by or addressed to women were dismissed by professionals as "popular" and therefore unworthy. One of the best known of these books was J. P. Nichol's *View of the Architecture of the Heavens* [1839], subtitled "A Series of Letters to a Lady." Among others, George Eliot was smitten by Nichol's work; she said it freed her imagination to "behold . . . floating worlds.")[5]

Another reason for the sun's popularity among amateurs was the social opportunity provided by eclipse expeditions. Various clubs for amateur skygazers, from the Liverpool Astronomical Society to the British Astronomical Association, sponsored journeys to various hilltops to observe and record solar eclipses. Walter Maunder, a Fellow of the Royal Astronomical Society, was instrumental in starting the BAA in 1890, in part because the Royal Astronomical Society's exclusion of women so offended him (though as we've seen, women's activities at BAA meetings were still highly restricted).

Physical astronomy was a field in which amateurs and professionals could happily meet. Professionals had access to sophisticated equipment, while the amateurs had a large rank and file and an eagerness to watch the sky for hours on end. It was at meetings of the BAA that the enthusiastic amateur Mary Acworth Orr met the budding young professional John Evershed. The sun—a dollop of Dante's Divine Light—would spark their courtship as they traveled to Norway and Algiers chasing eclipses.

7

Romantics

In late July 1896, Mary and John joined fifty-eight others, all of whom paid their own way—following the "admirable Pickwickian principle," said Walter Maunder—on an expedition sponsored by the British Astronomical Association to Vadsø Island, Norway, to see the total eclipse of August 9.[1]

Mary had read of ancient legends from around the world explaining solar eclipses as battles between the sun and the spirits of darkness, or the result of a sky-dragon devouring daylight in anger. Prominences glimpsed with the naked eye during total eclipses were rare and almost always misunderstood. In the course of her studies, Mary learned that during the May 12, 1706, eclipse, a British sea captain named Stannyan had recorded in his log "a blood-red streak" seen shooting from the sun's left limb, or edge.[2] It lasted for several seconds, he said. Of that same incident, Isaac Newton's friend Facio Duilier had written, "The clouds [of the 'solar corona'] became red, and then a pale violet. There was seen, during the whole time of the total immersion, a whiteness which did seem to break out from behind the moon to encompass it on all sides equally . . . [like a] halo."[3]

Balancing the rather mystical expectations inspired by these accounts were the practical preparations Mary and her comrades were forced to make for the trip. Travel guides published by eclipse sponsors advised, "Bring heavy blankets or rugs, or both, as the nights are extremely cold. . . . [Bring also] a flask of brandy in case the water is unwholesome."[4] Crates of food and remedies for animal bites, sunstroke, and cholera were loaded onto the steamers.

Opportunities for travel were improving rapidly just as Mary began moving around the world. Steamships and railroads converged, increasing the scale and speed of conveyance. Certainly, Mary had an easier time traversing dark woods than Dante did in his exile. In 1867, the year of her birth, one British travel agency proclaimed, in an ad, "Facilities in travel, more wonderful than the dreams of ancient poets, await the modern voyager, annihilating difficulties of time and circumstance, smoothing his path in the wildest regions, making his journey a mere question of time and money."[5]

Tilbury, England, Saturday, July 25, 1896: Who *was* the assured young lady stepping aboard the *Norse King* that day? She was listed in the ship's manifest as M. A. Orr. At twenty-nine, she was already an experienced astronomical observer, comfortable with technical equipment. She was the author of a book on the southern stars, their positions and their legends. She was a scientist and a poet. In her early writings—like the piece she would soon prepare on the Aborigines—she channeled the spirit of the Romantics, who, along with steamships and trains, flourished in nineteenth-century Britain.

Consciously or unconsciously, Mary linked herself to the tradition of William Herschel. Herschel practiced a form of "maverick" science, according to the historian Stuart Clark: "Instead of concerning himself with measuring the position of the stars to construct more reliable navigation tables, Herschel concentrated on discovery. This break with traditional [methods] . . . had a grand goal: to discern the complex interplay of the heavenly objects by incessant observation. It was a sentiment that resonated with the burgeoning romantic literary movement of the time, [which] believed that to see and experience a thing was a major part of coming to know that thing."[6]

In large measure, the poetic side of Mary had found Italy and Dante *because* of the Romantic movement. The critic Stephen Hebron writes, "From 1796, when Napoleon first crossed the Alps with his army, until his final defeat at Waterloo in 1815, the British were effectively cut off from Italy. Their knowledge of the country and its culture was therefore largely confined to what they found in literature and art."[7] After Napoleon's defeat, when Italy opened up again, British travelers discovered and fell in love with the country. The painter J. M. W. Turner was among those who responded profoundly to Italy's landscape and light. Byron and Shelley spent much time there, learning the language. Byron's affair with Teresa Guiccioli, the wife of one of Ravenna's leading citizens, immersed him in Italian life and politics. In 1818, Henry Cary published an English translation of *The Divine Comedy*. Keats and Coleridge praised the poem, and William Blake made a series of illustrations inspired by the Florentine bard.

One of the first poems Mary ever learned was Shelley's "Julian and Maddalo," set in Venice, evoking Italy's splendor and the spirit of the wandering Dante: "How beautiful is sunset, when the glow / Of Heaven descends upon a land like thee, / thou Paradise of exiles, Italy!"[8]

A few years prior to the Romantics' celebrations of the *Comedy*, William Herschel embarked upon an extraordinary series of lectures he hoped "would spark a grand discourse on the nature of the sun and the precise links it shared with the Earth," writes Clark. Herschel delivered most of these lectures in a splendid artistic setting: the Royal Society's Somerset House, on the banks of the Thames River. "Surrounded by . . . oil paintings . . . [and] wooden pews beneath chandeliers and an ornate paneled ceiling," Herschel traced a recent period of minimal sunspot activity.[9] Solar eruptions might well alter Earth's climate, he argued. (Sometime later, Mary's friend Walter Maunder would amplify Herschel's findings and explore their consequences.) The cultured atmosphere of Somerset House lent extra drama to Herschel's presentation and—Mary noted with interest—aided his effort to popularize science.

This, then, was the path, combining science, poetry, and history, that Mary followed with steady purpose from an early age. It led her confidently onto the *Norse King*'s barnacled ramp in the summer of 1896.

By that time, technical language was beginning to replace the "language of the sublime" in eclipse accounts, says Alex Pang, a science historian.[10] Measurements, procedures, and instrumentation dominated the vocabulary. Still, for many observers the sun remained an object of romance. The Pre-Raphaelite artist John Brett and the painter and illustrator Henry Holiday gave lessons at parties sponsored by the Royal Astronomical Society on how to capture, with pencils, the solar corona during an eclipse. Was the corona concentric, evenly diffused around the sun's disc? Was it composed of beams or luminous patches?

In some ways, the artist's eye remained superior to that of the technician. Pang says that, in eclipse photographs, "the tremendous differences in brightness between the inner and outer corona meant that one part of the eclipse could be recorded only at the expense of another."[11] Dark filters were required, distorting the image. A Romantic might put it this way: a photograph can't distinguish body from soul. Only an artist can do that, a painter or an ancient poet.

(It wouldn't be long before the nature of observation changed, as photography and spectroscopy came to dominate solar astronomy. In 1870, a British amateur, J. J. Aubertin, wrote, "Marvelous [astronomical] effects cannot of course be dwelt upon and enjoyed by scientific men for their whole attention must be engrossed by their observations and their instruments.")[12]

Mary would become proficient with cameras and spectroscopes during her years as a solar researcher in Kodaikanal,

India, but she would never lose her Romantic sensibility. The eclipse expeditions in which she participated were inherently romantic affairs in spite of all the equipment, involving the thrill of travel, the hunt for the perfect observing conditions. Watching through a telescope as the edge of the moon first made contact with the solar disc, she may have felt like Dante entering the substance of the moon in *Paradiso* II, or like the Pilgrim being transported in a flash of light from one heavenly sphere to another.

Mary had read scads of accounts by previous eclipse-watchers. They were crucial to her development as a thinker and as an observer. Aboard the *Norse King*, anticipating her own adventure, she vicariously relived incidents recorded by witnesses of the 1860 and 1871 eclipses in particular: "The clouds began to look dark and threatening, and appeared to lower . . . towards the earth, while the parts of blue sky gradually changed to a deep somber purple," one observer had written; strange colors "gave a ghostly illumination of the landscape"—an "ashy gray colour," wrote another, casting on people "an unearthly cadaverous aspect," making them look like "denizens of another world, so livid did their faces appear."[13] For Mary, this was an image straight out of *Inferno*, a world of living shades.

She tried to imagine the adrenalin flow released by the sight of a solar eclipse, which, she believed, must make these rare occasions into oddly spiritual experiences. As one observer she came across wrote, "Up to the instant of totality, the motion of the heavenly bodies had been so rapid that fancy almost created a sound; you fancied you heard the whirl of the moon rushing on through space." But then, as the moon's face obscured the sun's, an amazing stillness seized the sky, and the white corona blossomed: "As every eye watched eagerly the small glitter and dazzle of expiring sunlight, it is suddenly transformed into an indescribably beautiful halo . . . a white radiating glory . . . bearing a striking resemblance to the light which painters draw around the heads of saints."[14]

Ecstasy instantly gave way to emptiness as sunlight returned and the Earth resumed its daily form, bereft now of the radiant halo. The hollowness was especially profound if the observing conditions had not been optimal. Mary was quite moved by a comment of Dr. A. A. Common, president of the Royal Astronomical Society: it was as if "something had gone wrong, for . . . the silence [was] very oppressive." An amateur observer noted of a clouded-over eclipse: "As the light returned it showed a very disconsolate group. We could hardly look in each other's faces, and I am sure no one just then could find voice to speak."[15]

Whatever lay ahead of Mary as she stood unsteadily on the *Norse King*'s swaying deck listening to the steamer's engine blasts, feeling sea-mist sprinkle her face, squinting from beneath her hat into the midafternoon sun, she was eager to leave Tilbury on the 1,800-mile voyage to Vadsø Island, for an event forecasters predicted would last only 106 seconds.

On the first night out, rough seas pitched her stomach into trouble, and she couldn't face the following morning's breakfast of porridge, fried fish, grilled steak, liver and bacon, minced collops, buttered eggs on toast, boiled eggs, and curry. By noon the next day, when the ship made temporary landfall at Stavanger in southern Norway, she felt steady again.

One of the most prominent figures aboard the *Norse King* was Sir Robert Ball, who had been appointed the Royal Astronomer of Ireland in 1874 and who served as a professor at the University of Dublin at Dunsink Observatory. Gregarious, orotund, and raucous, he was a bit of a blowhard in Mary's view—an opinion John Evershed shared with her, a point of contact between them. For one thing, Sir Robert and his son, traveling with him, hobnobbed on deck only with other persons of note, such as Dr. Common, and mostly ignored the noisy amateurs from the British Astronomical Association. "Many ladies interested in science" seemed to be "found in the party," Sir Robert's son noted dryly, and rather dismissively, to his father one day.[16] Each afternoon, in the ship's saloon, Sir Robert lectured the amateurs on the upcoming eclipse. "Why did we not stay in London" for the event, he asked rhetorically one day. The Creator had made a mistake and not set the sun in orbit around Great Britain. The truth was, "as our American friends sometimes remind us," he said on another occasion, "England itself is only an inconsiderable patch of Europe, and Europe itself is only a very small fraction of the whole area of the earth. . . . There is only a very limited area of the earth from which it would be possible to see a total eclipse."[17]

Mary *did* appreciate Sir Robert's final lecture right before the ship landed at Vadsø Island. "Even if the fates are unpropitious, I shall regard this expedition as a rare and interesting experience," he intoned. "We have seen . . . beautiful fjords. We have learnt to know and love one another."[18] By that time, Mary was spending a good deal of time in John's quiet company. They marveled together at the snowfields and glaciers, the bare rocks, and the days' gradual lengthening.

The ship anchored briefly at Harstad, in the Lofoten Islands. There, the astronomers encountered Laplanders crouching in tents pitched in snowbanks, selling knives and other wares. Skittery reindeer huddled nearby on rocky escarpments. A few days later, when the travelers rounded the North Cape, a small musical group assembled on board to play the British and Norwegian national anthems. A salute was ordered, and the firing of guns. Just before midnight that night, the *Norse King* anchored at Vadsø, near a former whale processing station. A few goats clambered over an old stone quarry. The sun was four

Fig. 7. British Astronomical Association eclipse expedition in Vadsø, Norway, August 1896. Courtesy Alinari Archives, Getty Images.

degrees below the horizon, offering just enough light for the party to haul their instruments ashore. Using wooden carts provided by the Vadsø waterworks, they unloaded their tents and the rest of their provisions.

Over the next few days the usual maritime traffic—whalers sporting tiny crow's nests and swivel-guns, small Norwegian and Russian trading boats ferrying flour, wood, and dried cod—was joined by the grand arrivals of men-of-war and of yachts bearing dignitaries curious to see the eclipse, including the British M.P. Lord Charles Beresford and King Leopold II of Belgium (the man responsible for the hideous conditions in the Congo memorialized in Joseph Conrad's *Heart of Darkness*).

This far north the nights were short, an advantage with the instruments: temperatures and air moisture did not fluctuate to any great degree, so did not throw off the telescopes' fine calibrations. In the six days prior to the main event, however, the skies remained thickly overcast, plunging everyone into a funk. The goats wouldn't stay out of the food, and the island's predominant bird, the green cormorant, was too fat to fly properly, presenting an awkward sight that left everyone mildly depressed. Sir Robert, in the habit of wearing buffalo coats in the cold, bored people with incessant complaints about the disappearance of the buffalo from the American plains—future astronomers will be severely disadvantaged by the lack of warm buffalo hides, he said. Spirits cratered further when a midshipman aboard another eclipse party's boat lost his footing aloft and fell to his death. Everyone in Vadsø, residents and visitors alike, attended the funeral.

The BAA party organized itself into small platoons of observers, each with specific duties during the eclipse: those drawing the corona, those taking photographs, those like John Evershed overseeing the spectroscopes. For the first time ever on an eclipse expedition, a coelostat was to be used, a plane mirror engineered to turn on an axis at half the rate of the Earth's rotation so that light from the sun would always reflect in the same direction.[19]

Smoked glass lenses had been inserted into each of the optical instruments for eye protection. People without telescopes or binoculars were asked to record data or to work as general assistants. Mary was in fierce demand as a timekeeper, for both her precision and her kindly disposition. Some members of the party thought it a good idea to blindfold observers just before totality to increase the sensitivity of their eyes, but this practice was not generally supported.

First contact—when the moon's edge appeared to touch the sun's disc—was expected at four o'clock in the morning on

August 9. A bugle call awakened the party at two, and everyone left their tents to assume their positions. The sky was entirely overcast. Now and then, as the eclipse commenced, a thin crescent of sun could be glimpsed through dense, slow-moving clouds, but there would be, this day, no spectroscopes or spectacular coronal photographs. Later, a reporter for a paper called the *Daily Graphic* wrote:

> The light faded from the sky and land and sea. A feeling of apprehension and awe fell upon all. The birds flew home with a shrill bell-note of alarm, the island goats—throughout the week the irrepressible investigators of every instrument, and the chief enemy against whom our watchmen had to guard—sought refuge in a hollow of the rock. . . . The approaching shadow . . . came with appalling swiftness . . . as if . . . a great crape veil were being drawn over the world. . . . The clouds were of a supernatural blackness, or rather of the deepest indigo purple. Down on the horizon and here and there in little rifts the light was amber, and right below the sun a short rift of some two or three degrees above the horizon was even red. . . . Then . . . the veil was seen to pass away as if snatched rather than drawn, and suddenly the light returned.[20]

In silence, the disappointed group packed its instruments in sleek black boxes resembling coffins. People did not look at each other, and they faced away from the sun as if it had done them, and would continue to do them, irreparable harm. Mary sensed a feeling of collective shame at their failure—reminding her once more of Dante's penitential shades.

Later, Walter Maunder would say, "We did what we could, we stood and waited. We may comfort ourselves with the reflection that, 'They also serve who only stand and wait.'"[21]

His wife, Annie, a hunched creature known to her intimates as Mouse and soon to become one of Mary's best friends, nicely framed the adventure: "Those scientific fruits which had been hoped [for]" from the day were "scuttled." However, the BAA had proved it could mount a well-organized trip, and the excursion was "most successful" at cementing bonds among the observers.[22]

We may imagine that among the things John Evershed observed were Mary's demure smile and her dark eyebrows, often quizzically raised, which gave her a delighted but slightly melancholy look under her curly brown hair. A keen observer, Mary must have noted John's square jaw, high forehead, and gentle mouth, held tight. He was tall and his build was slight, reflecting a mild delicacy of constitution, but he had an astonishing stamina for craning his head to the sky hour after cooling hour.

In the early days of the *Norse King*'s return journey, danger and social excitement mixed to bring John and Mary closer together in a small whirlwind of emotional intensity. At Tromsø, in far northern Norway, the ship ran aground due to an engineer's error and sank into a muddy shore at low tide. Word spread among the passengers that they might drown as the waters rose. One man had packed a collapsible boat in his luggage; frightened by rumors of imminent peril, he inflated it and jumped overboard with it. Within a few hours, the crew had calmed most of the passengers, assuring them that the ship would shortly be freed. It was decided to go ahead with a planned evening concert in the saloon. At the concert, Sir Robert nervously joked that a gentleman who had been scheduled to sing "Rocked in the Cradle of the Deep" should offer a more appropriate tune, "My Lodging Is on the Cold Ground." Mary felt her tensions let go as she laughed along with John and the others.[23]

The fact that Sir Robert and his listeners had not actually managed to see the eclipse did not prevent him from lecturing about it every day as though all had been revealed to him. He became the butt of private jokes on board, but even managed to laugh at himself when some members of the traveling party staged a play one night titled *Teetotality*, featuring an absurdly comic character based on him.[24]

Mary, usually serious and not much given to social exuberance, found that she enjoyed the heady combination of scientific discovery and communal pleasures. She especially appreciated sharing these experiences with John, who began to emerge from his own habits of withdrawal in her presence as well.

Halfway down the Norwegian coast, the ship's crew received a telegram from a group of English vacationers that the *Norse King* should anchor near a certain hotel—the vacationers wanted to throw a fancy dress ball for the intrepid eclipse-observers. Somehow, the message got garbled, and the ship's revelers emerged onshore dressed for a costume party, in furry Laplander coats and polar-bear skins. They "looked like patients newly escaped from a lunatic asylum, who had ransacked the glass cases of an ethnological museum for articles of attire," said Sir Robert's son.[25] Mary and John did not dress up, but they joined the laughter and the dancing, as Norwegian villagers gathered outside the hotel's ballroom windows, gaping at the strange English scientists.

For reasons unknown, Mary did not accompany the BAA party to Talni, India (near modern-day Maharashtra), for the next eclipse, on January 22, 1898. The Maunders went with John Evershed. It was John's first visit to the country where, along with Mary, he would spend his most productive years as an astronomer and make his name as a solar researcher.

After several delays due to an outbreak of plague in India, the eclipse team boarded the RMS *Ballaarat* at Tilbury on December 8, 1897. They made their way to Marseilles, where

John observed a "great group of sunspots . . . very near the center of the disk."[26] Then to Brindisi. There, off the coast, they witnessed the zodiacal light, a "broad golden beam" streaking across the evening sky, a first for each of them—as Walter Maunder put it, a "new revelation, most striking."[27]

As the ship steamed across the Red Sea, John tested his spectroscope. His cabin was on the port side, facing east. He attached his instrument to the underside of his cabinmate's berth and aimed it out the porthole at the sun. The results were "very satisfactory," John reported, and Maunder wrote, "The trifling inconvenience which this otherwise excellent arrangement caused Mr. Thwaites [John's cabinmate] was, of course, cheerfully borne in the interests of science."[28]

The ship traveled due south. The group was astonished at how quickly the Pole Star sank each evening. Soon, the "new stranger southern constellations rushed up from the under-world," Maunder observed.[29] John retrieved from his pack Mary's book on the southern stars and referred to it faithfully, appreciating her thoroughness. The round blue maps reminded him of her large, startling eyes, he wrote later. He sat on the hard mattress in his berth, missing her smile.

The Southern Cross became prominent when the ship reached Bab-el-Mandeb, a strait located between the Arabian Peninsula and the eastern horn of Africa, a place known as Exile's Gate.

The team landed in India on January 3. By train they crossed the great central plain, from Bombay up the Ghats, through "flat, bare, monotonous, and above all dusty" terrain, Maunder wrote.[30] Eighteen hours later, with the help of camels and coolies, they made camp in a cleared cotton field surrounded by tamarind and mango trees, near the village of Talni, a "very ordinary" settlement of "small mud huts."[31] Their tents were roofed with bamboo matting, constantly fanned by the coolies using palm leaves in the middle of the day, when temperatures reached ninety degrees in the shade. Great earthen vessels formed the water cistern. The team mounted its telescopes on brick piers. John dug a pit behind his tent for photographic chemical waste. In this pit one day, just after teatime, he nearly stumbled over a two-foot venomous snake. The encounter shook him far more than the mournful howls of the jackals each night. The coolies informed "sah-ib" he was lucky, indeed, not to have been poisoned. These were very tricky snakes, they said. Every six months, they switch their tails with their heads, so one can never tell which direction they are moving.

One evening Maunder mused to John that on cold, clear nights, it was easy to "go back to childish fancy, or to medieval thought" and see the sky as a vault overhead.[32]

On eclipse day, the instant the moon's edge touched the sun, the temperature dropped. Maunder remarked that India would be "an ideal land if its sun could be perpetually half eclipsed."[33] As totality drew near, John watched for the coronal flash—the moment when the sun's upper atmosphere became visible, in a sliver of light becoming a burst of radiance along the edge of the sun's blackened disc. Observed through a slit spectroscope, this moment appeared as bright as a Roman candle, filling the whole field of view with dazzling lines too numerous to count. John exposed his spectrograph plate. The day went dark for a full two minutes. The usual scuttling sounds of lizards in the leaves underfoot instantly ceased. From the nearby village, a ghostly, frightened wail rose into the trees. Then: shouts of joy when the sun at last emerged like a "bursting shell."[34]

That night, the villagers treated the eclipse team to a lavish feast, to fireworks and dancing. They draped garlands of flowers around the Englishmen's necks. Stories reached the party from across India—how the sun's disappearance had prompted some state governors to release prisoners in a show of divine forgiveness, bestowing money and clothes upon them; how hundreds of people had taken holy dips in the Ganges as the moon extinguished the light. The team's photographs and spectroscopic plates were generally successful and of great potential interest in examining the sun's corona. Annie—Mouse—Walter's wife, got the most exciting results using a short-focus camera featuring a Dallmeyer lens. During totality, she managed to capture a streamer emanating from the corona in a "rod-like" fashion. Later, it was measured to be ten million kilometers, by far the longest extension yet recorded of the sun's corona. Mouse's picture bested professional images obtained by the world's state-of-the-art equipment. At a Royal Society party, Agnes Clerke, a BAA member, remarked, "Mrs. Maunder with her tiny lens has beaten all the big instruments."[35] Eagerly, the association anticipated the next eclipse.

On May 28, 1900, the BAA's plans to send a large party to Algiers, where the eclipse would be total, fell through when the ship the group had chartered from the Royal Mail Stream Packet Company was requisitioned by the British government.

Fig. 8. Drawing of Annie Maunder's coronal streamers by W. H. Wesley. From *Knowledge*, May 1898.

The government needed a vessel to run unspecified cargo to and from South Africa.

We know from a rather dry paper John delivered to the Royal Astronomical Society on July 16 that he and his brother Harry made it to Algiers and photographed the sun's flash spectrum from the nearby countryside. Barbara Reynolds, a noted Dante scholar who came into possession of Mary's papers after her death in 1949, writes that Mary also reached Algiers. The papers reveal that she traveled with four other women from the BAA, lugging with her a three-inch refracting telescope.[36] The women commandeered the rooftop of the British consul's villa as an observation post. There, they set up a metronome to beat the seconds of totality—the Music of the Spheres—and watched the sky turn purple, assuming a texture somewhere between a sunset and a thunderstorm. The wind turned cold. Through the 'scope, milky crimson flames jumped around the sun.

John spent time with Mary in Algiers before setting off into the country, and he probably rejoined her group once the eclipse was over. His mission had been funded by a government body whose title might, in an ideal world, describe *all* bureaucratic entities—the Permanent Eclipse Committee. Even though John and Mary were miles apart at the precise moment the moon approached the sun, they had journeyed far together with a common purpose. The shared adventure was heady, perhaps even transcendent, as the name of John's endeavor suggested: The Expedition to the South Limit of Totality.

How can one ever return from such as that? A lifetime romance had begun.

8

Prisms

John Evershed—Mary called him Jack—was born at Gomshall in Surrey in 1864. He came from a long line of farmers who trusted the stars' seasonal signs for planting and harvesting. His maternal great-grandfather ran a shop in Portsmouth selling nautical charts and almanacs. Of John's six siblings, two of his brothers would distinguish themselves as makers of precision instruments—Sydney in the field of brewing and Harry in agriculture. From an early age, John shared their mechanical aptitude. Initially educated by private tutors and teachers at a Unitarian preparatory school, John eventually attended classes at Croydon, where, he said, "by a lucky accident I passed the only examination I have ever been subjected to."[1]

Physically slight, he always tired easily, but he was stubborn, stoic, and, like his brothers, driven to succeed. As a boy, he was enchanted by a partial eclipse of the sun, which he had seen through a smoked glass lens in a local doctor's telescope. In the library, he studied eclipse records, such as the Syrian account of 1375 B.C.: "The sun was put to shame and went down in daytime."[2] Astronomers were romantic figures for him. When he was six, he had seen pictures in an illustrated broadsheet of Paris under a barrage of German shells, and read how the French astronomer Pierre Janssen escaped the city in a hot-air balloon to observe a total solar eclipse.

Through his eldest brother, a student at London's School of the Mines, he met chemist and entomologist Raphael Meldola, a Fellow of the Royal Society, who in turn introduced John to

Fig. 9. John Evershed, circa 1920. Courtesy Science Source.

Charles Darwin. John's romance with science grew. By the time he was thirteen, he was making his own telescopes using discarded opera glasses and toy prisms. Through the prisms he saw the solar spectrum, a glimpse into a hidden world that would determine his life's path. He especially enjoyed sharing his discoveries with his sister Kate. All his life, she would passionately support his experiments.

For a while he worked for a chemical manufacturer, but by 1890 his interest in the sun had led him to establish his own observatory at Kenley in Surrey, where he built a spectroscope, set up an eighteen-inch reflecting telescope, and began to photograph solar activity. (As a sideline, he also took brilliant color photos of butterflies.) Formally, his training in astronomy was not much more extensive than Mary's—wryly, he referred to himself as an "irresponsible amateur"—but a man was expected to pursue a career.[3]

John and Mary lived apart for the five years following their Algiers expeditions, but they saw each other often. John returned to his observatory in Kenley. Mary spent this period in nearby Frimley, Surrey. She set up her own observatory with the refractor she had taken to Algiers and focused her attention on the moon and variable stars (stars whose light output changes because of eclipses, unusual rotation, or physical shifts such as swelling or shrinking). Mary read more about the history of astronomy. She remained active in the BAA, one of whose members, Agnes Clerke, twenty-five years Mary's senior, became a cherished friend. In addition to being a skilled astronomer, Agnes was a renowned historian and a fellow admirer of Dante. (She had also, as a child, taken piano lessons in Dublin from the women on whom James Joyce based the spinster aunts in his short story "The Dead.") As a young woman, Agnes had lived in Florence for seven years, and Mary loved to compare memories with her of Dante's youthful haunts.

In 1904, Mary published a paper, "Variable Stars of Long Period," in the *Journal of the British Astronomical Association*. Her steady work shows a desire to pursue a professional career—and perhaps indicates an assertion of independence, as well as a friendly competition with John, who was busy on his solar studies just down the road.

Edmund T. Whittaker, a mathematician at Cambridge and secretary of the Royal Astronomical Society, had married a cousin of Mary's. Early in 1906, he was appointed professor of astronomy at Trinity College Dublin and director of Dunsink Observatory. He planned an intensive study of red stars, many of which were variables, using photographs taken with a fifteen-inch reflecting telescope. Mary's knowledge of variables made her useful to him; he invited her to Ireland to live with his family in the observatory residence. Unfortunately, he told her, she would not be paid for her work; she could only come as a volunteer.

Around this time—early '06—John received an invitation to become assistant director of the Kodaikanal Observatory in India's Dindigul District. The clear flash spectra he had obtained at the Talni eclipse in 1898 had impressed H. H. Turner, then the Astronomer Royal, and his Algerian observations of the sun's continuous emission lines had caught the eye of Sir William Huggins, a stellar spectroscopist and president of the Royal Society. When, in 1904, the government of India sanctioned the appointment of a European assistant to C. Michie Smith, the director at Kodai, Huggins enthusiastically recommended John.

Suddenly, John and Mary's time together, along with their productive rivalry, was ending. Although the Dunsink opportunity didn't appeal to her, the move would give her access to fine equipment. Professionally, John could hardly refuse the opening he'd been offered. He seems to have been stunned by the possibility that he might lose Mary. He asked her to marry him and come with him to the Palani Hills.

She had little time to sort her contradictory urges. To advance as an astronomer, she could work unpaid as her cousin-in-law's "computer" or be the trailing spouse of an observatory's assistant director. She knew that computers had made stirring contributions to the world of observational astronomy—most recently, she had read about the work at Harvard, where computers had compiled information for listing the spectral types of nine thousand stars. She also knew that many amateurs would leap at the chance to be at Dunsink Observatory. She remembered Sir Robert Ball telling stories aboard the *Norse King*—at Dunsink, he said, he had once received a letter from a grocer's assistant begging for a position at the facility: "I am a grocer's assistant, but my spirit is above the selling of sugar and tea and longs for communication with the skies," Sir Robert quoted the letter (no doubt embellishing its contents in the telling). "I pass many a sleepless night yearning for the sympathy of the planets. As I weigh out two pennyworth of figs in the balance I think of Libra, the constellation, and long to soar aloft with the celestials."[4] But Mary would lose her independence at Dunsink. Would the same be true at Kodaikanal?

The history of nineteenth-century British astronomy is replete with examples of husband-and-wife teams, such as Walter and Annie Maunder and William and Margaret Huggins, whose distaff side failed to earn equal credit for research and publishing, despite clear evidence of shared work.

Mary was thirty-nine years old—a "spinster" to the locals, though no such stigma would ever force her into marriage. It was unusual for a woman of her background to choose to remain single so long, and if she paid attention to such things, she might have felt pressure from her peers. But just as her father had insisted she be offered, as a girl, the same extensive education boys got, she always gave herself permission to consider alternate paths. Had she succumbed to peer pressure, she never would have written *An Easy Guide to Southern Stars*. She never would have boarded the *Norse King*.

Whatever she did, she would do it because she *wanted* to.

On September 4, 1906, at St. Mary's Parish Church, Cloughton, near Scarborough in Yorkshire, Mary Orr married John Evershed, a man who took her as seriously as her father had.

In one of the most celebrated passages in *Inferno*, in Canto V, deep in a place where no stars shine, Francesca da Rimini confesses to Dante that she and her lover Paolo were overcome by lust after a spell of "reading" together led their "eyes to meet." At first blush, John and Mary appear to have little in common with Paolo and Francesca, except that their love blossomed over "reading"—in their case, a common intellectual endeavor that involved deciphering patterns in the sky and poring over books and charts. Like Dante's pair, John and Mary managed to "move through the air," borne "forward" on powerful wills to work, and their wedding could have been blessed by Dante's words: "He / who rules the universe . . . / we . . . pray to Him to give you peace."[5]

9

The Notebook of the Sun

The Kodaikanal Observatory, only seven years old in 1906, was one of a series of English outposts established in southern India so that, as official word had it, "Posterity may be informed a thousand years hence of the period when mathematical sciences were first planted by British liberality in Asia."[1] A more pragmatic reason for constructing the observatories was to monitor the weather and to study monsoon patterns, so British properties here could be better protected in the future.

Kodaikanal was the first mountain observatory on the continent. As early as 1717, Isaac Newton had suggested that the Earth's atmospheric turbulence limited the usefulness of telescopes and that a partial remedy to this problem might be a "most serene and quiet air, such as may perhaps be found on the tops of the highest mountains above the grosser clouds."[2] By 1892, when the British Solar Physics Committee in India proposed establishing an observatory to study, among other phenomena, the effect of sunspot activity on rainfall, astronomers were taking Newton's idea quite seriously. The committee dispatched C. Michie Smith, an Englishman who had lived in-country since 1877, teaching physics at Madras Christian College, to find a suitable site. He recommended Kodaikanal, at an altitude of 7,687 feet, for its superior air quality. A foundation stone, carried by horseback up a winding bridle trail, was laid in 1895, but construction was delayed over objections by Norman Lockyer.

Lockyer was a British astronomer credited with discovering helium, along with the Frenchman Pierre Janssen, through spectroscopic observation of the 1868 solar eclipse in the tobacco fields of Guntur, forty miles north of the Bay of Bengal. Now, working with the Solar Physics Committee, Lockyer (who had never visited Kodai) complained that Michie Smith's planned structures were "too costly and too permanent." He said he believed in a more flexible observatory model and proposed, instead, "shanties" made of canvas and wood that could be moved if observing conditions changed.[3]

Michie Smith did not accept Lockyer's reasoning. In a letter to William Christie, the Astronomer Royal, on May 12, 1898, he wrote, "Of course I know that Lockyer's action is taken mostly out of spite against me."[4] Lockyer believed Michie Smith lacked proper training in solar physics. He thought Kavasji Dadabhai Naegamvala, director of an observatory in Poona, east of modern-day Mumbai, was the best solar physicist in the country. John Evershed had met Naegamvala on the eclipse expedition in 1898 and also admired his skills. Although Naegamvala was Lockyer's choice to be the first director of the new observatory, his work at Poona absorbed him, and Lockyer finally had to accept the committee's support of Michie Smith. At William Christie's suggestion, the men met to work out their differences. "My interview with Lockyer was trying but we both kept our tempers," Michie Smith reported to the Astronomer Royal.[5] It was agreed that permanent buildings would be constructed on the site. Michie Smith took up residence in February 1899 to oversee the work, after a thousand coolie-loads of books and equipment had been carted up the hill.

Before settling at Kodaikanal, John and Mary visited several observatories in the United States and Asia so John could get up-to-date on worldwide solar research. The couple left England for New York's Ellis Island on the Anchor liner *Columbia* in September 1906. John was immediately impressed by what a "good sailor"

Mary was.[6] He was susceptible to colds. His energy flattened quickly, though he never complained about conditions on the ship.

They visited Yerkes Observatory in Wisconsin, staying for a week in the house of the director, Edwin Frost, and they had a sad meeting at Harvard College with Edward Pickering, who was mourning the recent death of his wife. He confided to them that his interests in life were diminished, and the observatory's needs overwhelmed him. In mid-October, John and Mary traveled by train to California. They toured Lick Observatory, on Mount Hamilton in the Diablo Range just east of San Jose. At Mount Wilson, near Pasadena, John studied George Ellery Hale's spectroheliograph. The spectroheliograph is a device for scanning the sun's image through moving slits, which isolates a portion of the spectrum and lets the observer analyze the sun's gases. As John knew from the prisms he loved as a child, light breaks through a prism into the colors of the rainbow: red, orange, yellow, green, blue, shading into violet. The light of an approaching bright object will shift toward the high-frequency (high-energy) blue end of the spectrum—a blueshift, akin to a police siren getting louder as it comes closer. A redshift, indicating cooler conditions, identifies an object moving away from the viewer, like the siren's fade as it grows more distant.

With the development of more sophisticated spectroscopes, astronomers and physicists realized that by pinpointing different patterns of bright lines and the dark lines between them (known as Fraunhofer lines, after their discoverer, Joseph von Fraunhofer of Bavaria), they could determine the chemical compositions of distant objects. As Nigel Calder, an Einstein biographer, explains, "Lines at particular positions along the spectrum . . . are like stations on a radio tuning dial and these lines correspond to light of precise frequencies absorbed or emitted by particular kinds of atoms in the stars."[7] When aimed at the sun, a spectroheliograph can peel away solar layers.

At Mount Wilson, John examined Hale's instrument and made mental notes for constructing an improved version of it.

Beginning in mid-November, the Eversheds made their way to India by way of Japan, arriving in Kodaikanal on January 21, 1907. Coolies carried Mary and her luggage up the hill on a series of palanquins. She and John moved into the residential quarters recently built just for them at the top of Observatory Hill. The space was small but charmingly graced by a brick fireplace. Mary enlivened the interiors with plain, elegant linens. Eventually, she would arrange for a piano to be carried up and installed by the living room window. A butler served the couple breakfast every morning—tea, toast, eggs, rice, and salt-fish. Mary quickly adjusted to the local schedule and learned to take her lunch, called *tiffin*—generally, grilled chicken, mutton chops, curry, and rice—at two. She enjoyed doing her own shopping for dinners, buying cold meats and bread at a place called Creighton and Company, near a coolie ghat east of the Kodaikanal Basin, and rice at the Kodai Stores Depot.

Fig. 10. A residential building at Kodaikanal Observatory. Courtesy Science Museum Library / Science & Society Picture Library.

It took a while for her to get over the shortness of breath she experienced moving about at such a high altitude, and she kept experimenting with clothing, trying to find the right combination of comfort and protection from the sun. The hooped and hobbled skirts of the leisurely women living at the hill station were ridiculous, she thought, trapping them in a sort of paralysis in this rugged terrain, as were the woolen dresses ornamented with silver fern pollen, considered high fashion by the local British expats because the ferns, growing in ruffled gullies, were difficult to find. The women's husbands would gallantly clamber into steep *sholas* (small woods in mountain ravines) to gather the plants; the women would press them onto their dresses, leaving a delicate leafy imprint in the wool. Mary assured John she needed no such frippery. She stuck with long outfits made of linen or cotton, switching to heavier wraps from late October to early December, when the monsoons unleashed their unpleasant dampness.

John refused to wear "cholera belts," the flannel abdominal vests he was advised to don against intestinal disorders in the tropics. Quite rightly, he saw these silly garments as placebos that would have no effect on serious chronic inflammation.

The historian in Mary was delighted by competing local myths explaining how the Palani Hills came by their name. The most common story she heard was that "Palani" was a translation of the Sanskrit "Varahagiri," or "Boar Mountains," after the wild animals that seasonally damaged crops in the region. A variation of the story said that local sages, *rishis*, had once transformed into pigs twelve naughty children who had mocked the sages, and this accounted for the name.

Her favorite tale involved the god Shiva and his sons Subramaniya and Vighneshwara. Shiva promised to bestow a beautiful fruit ("palani") upon the one who could first circle the universe. Subramaniya mounted his peacock, Vahanam, and journeyed round the world. His brother knew that Shiva was himself the universe, and simply walked around his father. When Subramaniya returned from his travels and found himself the loser of the wager, he was inconsolable until his father told him, "Palan-ni," that is, "Thou art the fruit." The place where this touching exchange occurred was forever after called the Palani Hills.

Mary was surrounded from below by temples and farms (whose patterned fields recalled, for her, the agricultural glories of Virgil's *Georgics*), and above by Divine Light.

At year's end, the Kodaikanal Observatory director was required to file an annual report to the government of India. The report listed the preceding twelve months' primary astronomical achievements and assessed the state of the grounds, the equipment, and the staff.

In 1905, before the Eversheds arrived, C. Michie Smith reported that four observatory assistants of Indian descent worked with him, along with a Writer (a recordkeeper) and a Photographic Assistant. The "subordinate staff" consisted of a "book-binder and book-binder's boy, a mechanic, four peons and a boy peon for the dark room, and two lascars."[8] (Traditionally, "lascars" were Indian sailors. Michie Smith used the term in his reports to refer to general workmen whose primary job was to carry potable water uphill to the observatory.)

Michie Smith complained that the spectroheliograph building was nothing but trouble: "The roof continues to leak in various places. . . . For some unexplained reason, and in spite of the frequent reminders, only a small part of the [repair] work has been done [by Public Works]. The sliding roof which covers the sidereostat was nearly blown off the rails several times during the south-west monsoon."

The dome for the photoheliograph house had not arrived from the makers. The new seedlings he had planted as a potential firebreak (from California pines and seeds delivered from Lick Observatory) had died. In January, "a forest fire . . . swept round nearly half a mile of the boundary of the observatory grounds. . . . Some fifty blue gums were badly burned." The water well had gone dry nearly three months before, necessitating weekly, sometimes daily trips by the lascars to distant sources for drinking as well as photographic processing purposes.

Most vexing, the days' high temperatures, even at this lofty altitude, warped the telescope mirrors, and heat waves made it nearly impossible to obtain steady images of the sun. Michie Smith had tried cooling the rocky ground with vegetation, but his ground cover had withered. Inside, near the telescope, he had hung blankets and mats and wind screens, but they had done little good. To top things off, the spectroheliograph building was poorly placed. "Why the present site was chosen [for the building] is not known," Michie Smith wrote, "but it is too late to make a change now."

This was the tattered landscape into which John and Mary walked, and it would take a massive effort to turn things around.

On his first working morning, John climbed the spiral staircase inside the observatory's north dome. With the help of his Indian assistants, he pulled ropes to open and position the dome. He adjusted the clocks and the photographic plates. Later, he meticulously inspected Kodai's Cambridge Calcium-K spectroheliograph. It had never performed adequately. In time, he would refine it and construct the finest spectroheliograph of its kind, surpassing Hale's model at Mount Wilson.

To Mary, watching John putter in the mornings, the spectra resembled chips glinting in the mosaics of Florence and Ravenna. With the development of photographic and spectroscopic astronomy, scientists routinely referred to "mosaic images" of the sky, layering multiple pictures ("tiles") over one another or placing them side by side to achieve a greater observational field than single images could ever reveal. Mary was intrigued by the fact that a mosaic image was, technically, a series of separate instants, merging to compose a single moment: this reminded her of Aquinas's description of the angelic consciousness, balancing the sun and the moon to illustrate how angels could know all things—past, present, and future—at once.

(Recently, researchers have used spectroscopes to analyze Italian mosaics in some of the places Dante loved. Among their discoveries: ancient blue tesserae were produced by synthesizing cobalt ingots, ground finely then sintered with liquid enhanced by tin oxide; reddish coloration was obtained through a dispersion of copper and gold—the humble minerals of the Earth, the residue of stars, refined to *evoke* the stars.)[9]

Kodai's spectra were the stained-glass windows of a private cathedral in which Mary imagined communing with her poet. Meanwhile, John had little time for lyrical flights. Within a few months of his arrival at the observatory he was capturing the first-ever ultraviolet spectra of a comet—Comet Daniel, which had appeared in the sky along with the southwest monsoon. He noticed unusual tail-bands behind the comet's nucleus and determined that they were caused by carbon monoxide ions (ions are atoms or molecules that have acquired a charge by losing or gaining electrons). Following John's discovery, other astronomers began to more closely observe and document the complexities of comet tails.

With Mary's able help, he began a program of photographing sunspots and solar prominences, an uninterrupted collection of observations that continues to this day and forms a unique record of solar activities.

At the time, sunspots confounded astronomers. Eventually, Hale would prove them to be centers of strong magnetic fields on the sun, but for now John agreed with his friends Walter and

Fig. 11. A solar prominence photographed by John Evershed at Kodaikanal, May 26, 1916. Courtesy World History Archive / Alamy Stock Photo.

Annie Maunder that "we may never know" what sunspots are, "for their cause seems to lie deep down in the sun itself." Do sunspots, like whales in the ocean, swim just below the sun's surface, then breach the plain of fire? We can "know . . . a whale [again] if it has a harpoon driven into it," wrote the Maunders, "but how are we to know a Sunspot when it emerges again from the solar depths?"[10]

"The Eversheds worked in a funny time in astrophysics, in terms of the sun," reflects Brandon Brown, a physicist and science historian. "Before quantum mechanics matured, people had no idea why the sun didn't just collapse on itself."[11]

Mary was aware of her blind spots. On some days, she felt she had not progressed much beyond the buffoon in Tolstoy's fiction who asked, "Which is more useful, the sun or the moon? The moon is more useful, since it gives us its light during the night, when it is dark, whereas the sun shines only in the daytime, when it is light anyway."[12] Had modern science really advanced that far since Xenophanes of Colophon, who thought the sun an amalgam of flaming particles recombining every morning? Or Anaxagoras, who called the sun a burning stone? Plato, claiming it was igneous? Aristotle, insisting ether composed its core?

Using Kodaikanal's equipment, Mary furthered her education—with her husband's help—the way she'd always done: by teaching herself. Along with daily spectra, she kept a notebook of the sun, a record of close observations by which she would come to know the sun's being. Often as she worked, she considered Dante in exile, wandering the Tuscan countryside, or wading through the marshes of Romagna, near Ravenna, burning logs for a fire. What happened when he warmed his hands? *This* she knew: solar light trapped as carbon in the wood was freeing itself. Without sunlight there would be no tree for Dante to burn. No Dark Wood, no Earthly Paradise reflecting the Light of Heaven.

She had to bring this thing within her grasp.

What *was* that fiery object shining through the six-inch telescope in Kodaikanal's north dome (a telescope still battered by its trip uphill several years ago on the raw shoulders of coolies braving tigers, cliffs, steep ravines)?

First, Mary knew, the sun was whatever we intended it to be—in that initial layer of seeing, the shallow layer that wasn't *seeing* at all. For instance, in the early-Christian West, it was heresy to say the heavens weren't perfect; naked-eye sunspot reports were swiftly suppressed or discredited. Not so in the East, where no Christian bias prevailed. On January 10, A.D. 357, a Chinese astrologer's log recorded: "Within the sun there was a black spot as large as a hen's egg."[13]

But in the West the sun retained God's purity. In the early 1600s, when the German priest Christoph Scheiner told a fellow cleric he had observed sunspots through his telescope, his colleague refused to trust him: "[You] must be seeing things. I have read my Aristotle from end to end many times and I can assure you that I have never found in it anything similar to what you mention. Go, my son, calm yourself, and be assured that what you take for spots on the sun are the faults of your glasses or your eyes."[14]

It was Johannes Kepler—even before Galileo was ready to launch a theory—who insisted that sunspots actually existed *on* the sun. He said they were like "stains we observe on red hot iron, or like slag or dross on the surface of molten metal."[15] By 1798, William Herschel was turning prisms toward the sun to refract its beams. Herschel called sunspots "dusky pores." He said the sun's faculae, its bright-white mottles, were like "the shriveled elevations upon a dried apple."[16]

In this way, little by little, across a stream of centuries, we learned to un-see before we could see. Standing at the six-inch, Mary understood that *some* scales had fallen from her eyes. Not all. The same is true of us today. Briefly, so we may fully grasp the Evershed's' challenges, here is what we have learned about the structure of the sun since Mary's time.[17]

First, no solid surface covers the sun. It is composed of numerous layers. The Core, the source of its energy, burns at a temperature of around 27 million degrees Fahrenheit. This intense heat frees the internal structures of the Core's atoms, breaking them into their constituent parts: protons, electrons, and neutrons. Neutrons have no electric charge—they interact minimally with their surroundings. But the protons, positively charged, and the electrons, negative, slam wildly into one another. In the process, neutrons and protons find new, more beneficial organizational arrangements, creating abundant energy nearby to work with.

The Core is enveloped by the Radiation Zone. The energy emanating from the fusion reactions at the sun's Core begins to travel up through this zone to the sun's outer regions. Here, the temperature is cooler than at the Core: some atoms remain intact. They are capable of absorbing the Core's wriggling energy and emitting it as new radiation. In the Radiation Zone, energy generated from below passes randomly, atom to atom. It takes approximately 170,000 years for the energy released in the Core to make its way out of the Radiation Zone.

The next layer is called the Convection Zone. The temperature is relatively cool here—3.5 million degrees Fahrenheit as opposed to nearly 9 million in the Radiation Zone. At this level, atoms absorb energy but because of the relative coolness they don't readily release it. Energy transfers happen slower, by *convection*—like bubbles in a pot of boiling water, the hotter material will rise, while the cooler stuff sinks to the bottom. As the hotter material reaches the top of the Convection Zone, it starts to cool. As it falls, it heats and reverses course once more, creating an enormous, constant roiling. Ultimately, the very hottest material moves through the Convection Zone, and fairly rapidly—in little more than a week—hauling energy with it.

The Photosphere, the next layer up, is what we, from our terrestrial vantage point, perceive as the sun's surface (though it is gaseous rather than solid). Traces of convection bubbles filter through the Photosphere, like swarming grains of rice. The temperature here is cool, about 9,980 degrees Fahrenheit, so the gas remains thin—thin enough for atoms to absorb and release energy. Most of the sunlight we receive on Earth is energy released by atoms in the Photosphere. It takes this light about eight minutes to travel 93 million miles to hit us—a fact that would intrigue Albert Einstein and nudge him toward the first of his breakthrough theories.

To recap: the time frame for all that massive energy cooked up in the Core? One hundred and seventy thousand years in the Radiation Zone, one week in the Convection Zone, and eight minutes from the Photosphere to the Earth. A miracle worthy of poetry.

Sunspots appear in the Photosphere. They look dark to us because they run cooler than the surrounding gas, averaging about 5,840 degrees Fahrenheit. Magnetic disturbances below the sun's surface infuse these spots, suppressing the heat beneath them. They are gargantuan, several times the size of Earth. They brew solar storms. Their life span can be as short as an hour or as long as many months, and their frequency levels vary in eleven-year cycles.

Mary's notebook of the sun was packed with drawings of the Chromosphere and sketches of the magnificent Corona. The Chromosphere is the realm of color, a layer of gas about 1,250 miles thick. Here, energy continues to be whipped about by radiation. Hydrogen atoms absorb energy from the Photosphere, most of this energy erupting as red light. Convective cells, much larger than those in the Convection Zone, dance throughout the Chromosphere, along with writhing flames shooting up thousands of miles. Right above the Chromosphere the temperature rises dramatically, from 35,540 degrees Fahrenheit to over 3.5 million degrees, possibly because of the sun's magnetic field (still not fully understood) or because of the spicules licking up from below. Then we reach the Corona, the sun's outermost layer stretching far into space, so called because of its crownlike appearance during a total solar eclipse. Particles from the Corona stream outward as "solar wind," reaching Earth's orbit. When the Earth's magnetic field traps electrons and protons from the solar wind, pulling them into Earth's atmosphere, they interact with the atoms there. The result is a massive release of energy, the Aurora Borealis.

The sun's magnetic field determines the Corona's shape. Sometimes the magnetic field emerges from the sun's lower regions then loops back down—visible to us through telescopes as dervishlike flames crackling all the way up into the Corona. Mary particularly loved cataloging solar prominences. They billowed and chuffed into space like the sails of ships torn loose.

Why *doesn't* the sun blow itself apart, as the Eversheds and their contemporaries thought it must? Essentially, the Core is too hot to burn. It seethes too much to be either a solid or a liquid. It is a gaseous soup of charged particles. The sun's serial motions are ruled by magnetism, made possible by convection, the bubbling and spinning of gas beneath the surface.

Eventually, the sun will consume its hydrogen, followed by its helium, triggering nuclear reactions. It will expand and then contract, losing much of its mass through solar wind. In five billion years, the sun will gutter out, like the cold candles in the old lighthouses on Mary's Cornwall coast.

We know what we know about the sun because pioneers such as John Evershed devoted their lives to its study. On January 7, 1909, he noticed on a spectrograph a radial movement of matter in the penumbral region of a sunspot—a wavelength shift indicating an outflow of gas. He searched for and confirmed the existence of similar motion in other sunspots. It became known as the Evershed Effect, a still-puzzling aspect of the sun's dynamics.

At a steady rate he published papers in the same dry prose that characterized his eclipse report to the Royal Astronomical Society. But as time went on, a curious quality crept into his writing. Was it . . . poetry?

Fig. 12. John and Mary reading mail together in India. Courtesy Science Source.

In one report, John switched suddenly from a rote discussion of "absorption hypotheses" to a meditation on sunspots' "smoke-like veils."[18] His 1910 remarks on Halley's Comet began with equations about the comet's "relative speed of approach" to the Earth, moved into praise of the tail's "magnificent spectacle," then into a description of the "square of Pegasus, which was filled" with the comet's "faint light."[19]

A 1917 report on the "results of prominence observations" includes this paragraph, remarkable for a scientific treatise:

> Day after day, sitting quietly in his own observatory, the astronomer can see the wonderfully varied forms of . . . huge flames, can measure their motions, and investigate

> their composition. Nothing can be more entrancing than to gaze at a great cloud, glowing as if with sunset colour, and full of intricate detail, which slowly changes, and presently grows dim and fades away, or to watch a number of fine sharp rays which shoot up, curl over, and disappear, while others take their place. In photographs, the colour and brilliance is missing, but the forms lose nothing of their strangeness and beauty.[20]

It is clear, in comparing this passage with John's earlier work, that this is not his writing. Before releasing the 1917 report to the public, he agreed to give his wife full co-authorship credit. With earlier papers, it is difficult to say how or to what degree collaboration occurred, but at the very least we see a meeting of minds, a fruitful conversation, a melding of poetry and science.

Following her article on variable stars, Mary published in the *Monthly Notices of the Royal Astronomical Society* a significant paper on solar flares—which owed much to her husband's observations with the spectroheliograph.[21] She was trying hard to find her place as an amateur with gifts as strong as those of the professionals around her. Her contributions to John's reports, combining history with detailed observations and a poetic flair, built on her earlier work and indicated a path her writing might follow.

In the meantime, she indulged the other great passion of her life: reading Dante. She recalled the pleasure she had taken in preparing *An Easy Guide to Southern Stars*, returning to the moment when the sky became fresh to her once more—an echo of *Purgatorio*'s final lines, when Dante sees the stars anew just as he is poised to enter Paradise.

10

The Gift of the Forest

It was a medieval sky Mary observed from her hilltop. Dante's understanding of astronomy was vitally influenced by ancient Greek texts—particularly the works of Aristotle—and so was India's, though the influence moved in both directions: the *Aryabhatiya*, India's first astronomical work attributed to a single author, Aryabhata I, written in A.D. 499, was translated into Arabic as *Zij al-Arbhar* in about 800.[1] (Historically, the region of the world currently split between Afghanistan, Pakistan, India, Nepal, Myanmar, and Sri Lanka was referred to, collectively, as the Indian Subcontinent. Unless indicating a specific location, I will use the name "India" in the following discussion.)

India's oldest literature, the texts composing the *Rig Veda*, forms the basis of the Hindu religion. Accurate dating of these texts has been elusive. Vedic culture is generally said to have flourished between 5000 and 1500 B.C. An addendum to the *Rig Veda* titled *Vedanga Jyotisha*, establishing astronomy as an essential subject of study ("Just like the combs of peacocks and the crest jewels of serpents, so does Jyotiṣa [astronomy] stand at the head of the auxiliaries of the Veda"), has been thought by some historians to be composed around 1400 B.C.[2] These scholars base their calculations on the *Vedanga Jyotisha*'s mention of the winter solstice in conjunction with the star group Sravistha (modern-day Alpha Delphini).[3] This suggests knowledge of the precession of the equinoxes, the discovery of which occurred around that time.

The Vedic texts make no distinction between astronomy and what we would call astrology—observing the sky as a means of deciphering divine revelations. The texts' observations are related in rigid metrical forms called *shlokas* (Sanskrit for "song").

Eclipses are attributed to two demons named Rahu and Ketu, who eat the sun and moon, respectively. Along with the Mayan civilization, the ancient Hindus appear to be among the earliest people whose sense of time extended beyond a few thousand years. For example, as experienced by the central Hindu creator-god Brahma, a single day or night is said to be 8.64 billion years long.

History lists Aryabhata I, born in A.D. 476, as the country's first true astronomer, ushering in Siddhantic (i.e., mathematical or computational) astronomy. Little is known of him. Scholars argue about where he was born, some insisting he lived in what is today Bihar in the north, others, citing references in his work, placing him in a region called Ashmaka in what is now the southern state of Kerala. What is certain is that a powerful school of astronomy and mathematics, based mostly on his writings, flourished in Kerala from the thirteenth to the nineteenth centuries. He immersed himself in Aristotle. In 499, his *Aryabhatiya* circulated as an invaluable book on navigation using the stars. If Kerala was his home, he may have witnessed the total solar eclipse of January 4, 493. Heavenly events were still thought to be divine revelations at that time; some historians speculate that the leaders of the caste system excommunicated Aryabhata for interpreting the eclipse in rational, rather than astrological, terms. In any case, he was a keen observer, accurately calculating the orbital periods of the visible planets. His watchful eye led him to wonder about a heliocentric system and to believe that the Earth rotated on its axis.

For centuries after Aryabhata's work became celebrated, Indian astronomy tasked itself with calculating planetary orbits and figuring algorithms from them, using sundials and water clocks. The first record of observatory construction occurs in the early eighteenth century with a man named Maharaja Sawai Jai Singh II of Jaipur. On orders from the Mughal emperor Muhammad Shah that time be measured more accurately, he built five observatories, called Jantar Mantars, beginning with one near Delhi in 1724, using mud and thick masonry. He fashioned the structures after palaces or grand temples. He constructed hollowed-out stone hemispheres latticed with crosswires to align star positions with certain markings on the rocks. Using a stone triangle, seventy feet tall, its hypoteneuse running parallel to the Earth's axis and aimed at the North Pole, he conducted precise terrestrial measurements. By now, astronomy had become a highly pragmatic pursuit, with applications for trade, navigation, and military adventurism.

Telescopes came into use in 1651, some forty years after Galileo turned his 'scope toward Italian starlight. India's first telescope users were French Jesuit priests such as Father Jean Richaud, who made a few astronomical observations in 1689 at Pondicherry, in southern India, for mapping purposes. With the advent of the British East India Company and its forays deeper into the continent in the late eighteenth century, the need for accurate geographical information grew more acute. Surveying instruments were at a premium—sextants, quadrants, clocks, telescopes. The first public observatory established outside of Europe was at Madras in 1786—the direct precursor of Kodaikanal.

By the time Mary and John settled in Kodai in 1907, Europe's medieval period was long finished, but India was only just beginning to emerge from its own long medieval cultural age. It was, according to historian S. Abid Husain, an era characterized by a robust Hinduism and a "vague national feeling in the political sense" untouched by any "fusion" with the West. In Husain's view, the Indian medieval mind was shaped by a geography of "plains or low plateaus, watered by big rivers," as well as a steady temperate climate, forging a "common history" and "economic unity" among the nation's peoples, in spite of lingering tribalism and linguistic parochialism. On the most primitive level, Indian culture has always been infused by an awareness of "vast area[s] of land steeped in the mysterious silence of a moonlight night in summer," Husain writes, a "feeling of solitude. . . . You feel as if there is nothing in the great infinite universe except you and the star-spangled heavens."[4]

Mary took daily walks for exercise, to appreciate the beauty of her surroundings, and to get to know the local villages. Strolling downhill from the observatory with her pith helmet fastened firmly on her head, she always felt herself plunged into the past. From the high plateau she wound around ancient granite shelves covered with grass, in and out of *sholas* in the mountain ravines, areas shot through with quartz folds and feldspar seams among clear, misty waterfalls. Mimosa and eucalyptus trees shaded her path. Green pears, coffee, and jackfruit. Crows called. And mynah birds. Scorpions scrabbled at her feet. She stepped carefully past streams where John liked to shuck his shoes and go crabbing on his rare days off.

If she was on her way to visit the Kukkal Cave, she'd cut through the village of Pumbari, home to four hundred people tending terraced hills and chickens. It overlooked blue-green forests rustling with buffalo. In the countryside, she'd immerse herself in a sensory blur: bamboo, banana, and sugar cane; banyan trees wide as stone huts; men in dhotis, women in yellow saris; shrines to Shankar and Rameshwar; lakes the blue of Arctic ice; sunlight tangled in trees; brown bodies, black hair; crickets and frogs; rumors of bandits ("You must carry a gun!"). In a marketplace, she'd rummage among seed-grinding stones and mustard-oil lamps; biscuits, boiled eggs, and tea; parakeets and swallows; quinces, guavas, grapes; chili peppers, lemons, and halva; pan, copra, bidi. Back in the *sholas*, she'd encounter lizards the size of a human foot, bees the size of British coins; armchairs sitting all alone in a field beneath mosquito netting, abandoned by some impatient Englishman; donkey tracks in the dirt; pipal trees; oxen and boars; a whiff of opium from a darkened village alleyway; swamis living on curds and leaves; near-naked men pedaling bicycles and playing harmoniums; bhajans, sitars, deer-skin drums; finger-cymbals clanging distantly among thick, dark groves of asoka trees; Ganesh the Elephant Deity, Hanuman the Monkey God.

And always, a strong sense of exile.

Deep in the woods she would come across dolmens and stone circles, some littered still with chipped funeral urns and fragments of blackened human bones. Ancient Tamil texts praised the god Murugan ("eternal youth"), a resident hill spirit born of Shiva and Parvati (in the yogic tradition, Shiva was a being responsible for both protection and destruction, Parvati was a Hindu goddess of love, devotion, and fertility—*parvata* is a Sanskrit word meaning "mountain").

In some more informal sects, Murugan was said to be the product of the sun and the moon, and he was often, in legend, accompanied by a peacock. Occasionally, from a distance, Mary would witness village dances in honor of this god: men garlanded with flowers, smeared with sheep's blood, moving slowly inside a circle of spears.

On her way back up the path to the observatory, she would leave the past behind. With each new terrace, the air cooled, and threats of malaria, cholera, typhoid, and dysentery lessened just a little: this was the reason English and American missionaries had established hill stations high above the sweltering plains.

The missionaries and the laborers that followed them here had never quite meshed. A popular slogan in the area captured

Fig. 13. John (seated) and Mary (at right) with unidentified friends in India. Courtesy Science Museum Library / Science & Society Picture Library.

the tension between them: "When Kodai Spiritual departs / The Kodai carnal season starts."[5]

Mary never felt at ease in the hill station proper. Ground was constantly being cleared for construction, stirring up dust—new housing (there were roughly one hundred and fifty houses in the station when the Eversheds arrived), new shops, a hospital. Mary and John rarely socialized at the English Club, established in 1887 to "cater to the social, recreational, literary, and other cultural needs" of expatriates, according to its advertising.[6] There, Michie Smith relaxed, riding his horse, Jerusalem, around the track. American, Scottish, Welsh, and Irish club members, as well as Britons, gathered to play billiards or badminton. It was the kind of exclusive place Mary had mostly avoided even back in England, though she attended an occasional tea-and-tennis party.

She did not particularly relish the mall, the central avenue with its big Anglican church, built to encourage British residents to feel at home. Something about the area—its severe architectural angles—reminded her of a private fortress. She was easily irritated by the grousing of fellow Britons. It was a common complaint that, although hill stations had been created as idyllic reserves for British citizens living abroad, growing service needs meant more Indians were moving into the neighborhoods, adversely changing them. Overcrowding and pollution had become serious problems.

Before arriving here, Mary, ever the historian, had studied up on the first European to visit Kodaikanal, a British surveyor named B. S. Wood. He came in 1821, followed by hunters and botanists. In 1834, American Protestants opened a mission nearby, in Madura (present-day Madurai). They relocated to Kodai after a cholera outbreak. By 1850, ten European bungalows sat among dozens of Indian huts at Kodaikanal. British and French civil servants planted coffee and tea, and a settler named Major Partridge of the Bombay Army imported Australian eucalyptus trees. Just five years before Mary and John arrived, the first sanitary water storage and piping systems were constructed near the observatory—established after a series of famines in India, which some solar observers thought were linked to patterns of sunspot appearances. In 1877, with hunger spreading in Madras, Norman Lockyer had suggested to Lord Salisbury, the secretary of state for India, that to safeguard English citizens abroad solar activity should be studied by keeping a daily photographic record of the sun. After all, Britain controlled more of the sun-on-Earth than any other nation—it was the Empire on Which the Sun Never Set. The photographic project began at Dehra Dun. It moved to Kodaikanal, along with the six-inch telescope from the old Madras observatory, when C. Michie Smith began his work at Kodai.

In *Paradiso*, in passages on the Heaven of the Sun, Dante acknowledges our stormy world and introduces the theme of universal harmony. In this part of the poem, he also laments the planet's poverty. Mary pondered these passages often. In her sunlit observatory, she was haunted by the cries of the poor from the hillside below—screams over scummy water, hunger, and disease.

Dante describes Thomas Aquinas circling the sun, in verses echoing the dirgelike rhythms of the Book of Lamentations. As Aquinas's spirit spins in space, he tells the tale of Francis. The saint marries Poverty, whom Aquinas describes as a widow, a living form of abandonment, despised and shunned by society. In choosing Lady Poverty as his bride, Francis restores her dignity. He expresses his empathy for the poor, mourning the world's sadness.

Dante's "espousal of mourning was lifelong," beginning with the death of Beatrice, writes the Dante scholar Ronald Martinez. Mourning became Dante's "vocation." Like the exiled Cain, he became a "fugitive and a vagabond" on the wind-whipped earth.[7] In Dante's story of Francis, Lady Poverty becomes physical death. Christ was her first husband; his crucifixion widowed her. When Francis dies, he is laid upon the ground—the very *essence* of Poverty: "From his lady's bosom [he] / chose to set forth . . . [His] corpse would have no other bier."[8]

In Paradise, God resurrects Francis, just as He restores Beatrice. Dante reunites with her in a glorious ceremony; together, they circle the heavenly spheres.

In Mary's walks in the woods and to villages near Kodaikanal, in her work with her husband on the hill, and in her meditations on Dante, she developed the firm conviction that the way to weather melancholy storms was to seek the center of heaven. As her writings made clear, she perceived her sunspot data as an ongoing poem, a record of daily lamentations.

The Murugan dances reminded Mary of the Aborigines she had encountered in New South Wales. Her mind turned again to the problem of astronomical origins. Shortly before moving to Kodaikanal, she had published an examination of astronomy in the Old Testament in a small scientific bulletin (later, this material would be incorporated into her Dante studies). She wrote:

> The Jews were forbidden to study and forecast the movement of the heavenly bodies, lest they be led away into star worship and star divination. So says the Talmud. Yet some knowledge of astronomy is necessary to a nation, and especially to her priests, for only by observation of the heavenly bodies can the dates of festivals be accurately fixed. How far did this knowledge extend with the Jews?
>
> The data are scanty and unfortunately just when we might expect to find light—namely in the ancient Jewish calendar—we are in the dark. The month was evidently lunar, from its Hebrew name and from frequent mention of festivals of the new moon; the year was as clearly solar, since the three great yearly festivals were all connected with the seasons. But

> a solar year does not contain a whole number of lunar months, and the problem of bringing the two into accord has taxed the skills of primitive astronomers of all nations. We cannot be sure how the Jews solved the problem.[9]

Biblical references to night and day perplexed Mary. In Exodus, she noted the curious phrase "between the two evenings." She speculated that evening may have been divided into two parts, beginning first at sunset when it remained light enough to work, continuing into a "second" phase coinciding with the rise of a crescent moon. The Babylonians had developed sundials by 700 B.C. The Hebrews should have known of them—but that couldn't be confirmed, reading the Bible. "If dials were used, there would surely be some mention of the divisions of the day more exact than 'in the heat of the day,' 'early in the morning,' etc.," Mary wrote.

Aside from a few constellations—Kesil and Kemah in the books of Amos and Job, for example (probably Orion and the Pleaides)—the Old Testament remained disappointingly starless. The "attempt to formulate a Hebrew cosmology does not appear . . . altogether successful," Mary concluded. "The truth is that the ancient Hebrews felt no intellectual need, as did the Greeks, to construct word schemes in order to explain natural phenomena. The universe was to them . . . simply the marvelous and inscrutable manifestation of one supreme power."

On the other hand, early Christians had a great flair for drama (thorny crowns! Crucifixion!): star divination fit their culture nicely. At Kodaikanal, late at night, resuming her obsession with history, Mary questioned the origin of the Star of Bethlehem. She was not content to write it off as a miracle sent by God for which there could be no natural explanation. And the Gospels were full of contradictions: the star was visible to the Magi but not to Herod; the star appeared in the east but the

Magi headed west to Judea. Mary figured the phenomenon had two possible sources: either it was an unusual configuration of known celestial objects (say, a conjunction of the moon and Venus) or it was a singular occurrence, a comet or a nova. (An additional problem was not knowing the actual season of Christ's birth. For their celebration, Christians had adopted the Roman Winter Solstice of the Unconquered Sun. In this, their calendar came from a sixth-century monk, Dionysius Exiguus, who had apparently miscalculated the reign of the Emperor Augustus. Christ was probably born a few years earlier than the calendar suggested.)

Chinese records for March–April, 5 B.C., noted that a "broom star"—that is, a comet—was visible in the morning sky for several weeks, but Mary thought this an unlikely explanation for the star. In both Eastern and Western traditions, comets had always been feared as harbingers of bad news, not greeted as causes for joy. Besides which, though rare, comets were not entirely uncommon.[10]

No novas—exploding suns—were recorded at this time, not even in China.

Was it possible that the "appearance" of a divine star was an abstract mathematical calculation rather than an actual stellar event? Astrologers of the time were well known for predicting rare conjunctions invisible to the naked eye. Once again, this might be a matter of un-seeing.

Mary walked out into the night, in the moon-cast shadow of the observatory's north dome. She tried to imagine herself back in time, viewing these very same stars without knowing their names, without recognizing the familiar shapes by which she had always identified them. Down the hill, in the dark, the droning of a sarod. She thrust her mind into the sky and pulled it back into her mind, imagining it as a fresh slate in which the Ptolemaic structure still lay far in the future.

11

The Scarcity of Wasps in Kashmir

Mary's marriage to John was primarily an intellectual partnership, animated by a shared love of adventure in spite of its hardships. They never had children. The correspondence and papers they left behind were mainly of a professional, rather than a personal, nature. Above all, they shared a passion for observing—not just the sun and the sky, but the landscapes and the wildlife around them, the various cultures they moved within as visitors and exiles.

Mary worried about John's hard work ethic. He wore himself down. He berated himself for the slightest slip-up or failure with equipment. If she mentioned she'd like to plant a garden—some lupines and larkspur, a few tidy chrysanthemums—right away, he was out digging her beds. He ordered an "English mamooty," a state-of-the-art earth-scooping tool, so he could gouge a bank near the residential bungalow for building a new spectrograph. "Iron rails and queer-looking stands and bits of things are gradually accumulating . . . and the new liquid prism is to be used in the spectrograph, which we hope will do great things," Mary wrote John's sister Kate. Mary fretted for John, sweating for hours in the sun, but she knew how important it was for him to do things right and to create a remarkable new observing tool. "It is his own," Mary said. That's what mattered.[1]

She talked him into buying a dog, a mutt from a market down the hill that she named Remus. Remus's evening walks were Mary's excuse for stealing John from his grueling routines, forcing him to relax.

Fig. 14. John working with astronomical equipment at Kodaikanal. Courtesy Science Source.

When C. Michie Smith retired and John became director of the Kodaikanal Observatory in 1911, the Eversheds' marriage evolved, more than ever, into a solid work-partnership. In addition to his research responsibilities, John became accountable for the operation's finances, upkeep of the equipment, and the well-being of his staff—Sivarama, Sitarama, Nagaraja, Balasundarum, Krishnaswamy, and Krishna Aiyar, along with Parthasarathy Devendran, whose descendants remain at Kodaikanal to this day, successive generations of sun-watchers.

John scheduled the observatory's activities around the wet season from July through November and the dry season from mid-December to May. He continued to grapple with leaking roofs, grass fires, a dry water well, electrical shorts, and a Public Works Department that didn't know the meaning of work. New instruments he'd ordered either arrived later than promised or were too heavy to be carted uphill by anemic mule teams. Often, equipment needed to be built from scratch. Overwhelmed by administrative and maintenance tasks, he asked Mary to conduct more daily work with the spectroheliograph and the cameras.

Sometimes John and Mary had to protect each other physically. The 1910 annual report noted that one night "a fire [was] lighted inside [the compound]—evidently maliciously . . . [and] destroyed 50 young trees." The bungalow was threatened. Every available staff member was engaged to fight the fire—the flames came very near the spectroheliograph house, leaving blackened soil and smoldering grasses that severely impeded the steady air necessary for accurate solar observations.[2]

Implicitly, Mary trusted her Indian assistants, but she also felt her outsider status. She had traveled the world. She had visited Dante's Italy. She had lived in Australia. She had toured America and Japan. Like any Western woman of her time, she carried biases, but by and large she understood cultural clashes and the nasty politics inherent in colonial enterprises. She knew the history of the British in India—they had been little more than pirates in the days of Queen Elizabeth I. She knew how England's conquest of the continent had become essential to its view of itself as an empire. She had encountered the attitude, so common among English civil servants abroad, that the British had a moral responsibility to uplift less civilized races; conversely, she had heard Indians speak of the "Northern peoples'" lack of imagination, their weak emotions and tardy responses to stimuli.[3] As a young woman, she had not been immune to British tabloids' lurid depictions of Indians as thugs committing bloody atrocities on defenseless infants and breastfeeding mothers.

Even now, around the English Club at the hill station, talk frequently turned to racial exclusivity, and the comments were not always aimed at the native population. Club members made strict distinctions between "white" Europeans—that is, wealthy families who could afford to send their children to private schools back in Britain—and those they called "country born": working-class Britons who had come to the colonies to work in commerce or to lay railroad tracks. They had, they said, a "touch of the old brush in them"; their "chi chi" (lower-class, rather singsong) accents often gave them away. Mixed-race or Anglo-Indians were shunned altogether.[4] Mary was aware that even among privileged families here, it was the boy children who received the educations back home, while the girls, whose instruction was not seen as important, were kept at the hill station, and tended to graduate from Presentation Convent High School in Kodaikanal.

In this swirl of fierce identity-pride, and following the observatory's arson incident, Mary became more conscious of anti-British sentiment in and around Kodai. Many local Hindus resented the fact that Christian theology had become a compulsory subject at the University of Madras. They chafed against a law allowing Christian converts to inherit private property from their Hindu ancestors. She read copies of English-language newspapers published by independence activists, *The Hindu* and *New India*, and she read the work of Tamil Nadu's beloved poet Subramanya Bharathiyar, until his odes to freedom were banned by the government and he fled to Pondicherry in exile.

Mary's occasional fears for her safety did not stop her from doing what she had always done—put on a pith helmet and hike into the fields. In addition to her notebook of the sun, she kept a record of wildflowers she observed in the hills: Traveler's Joy, Old Man's Beard, Buttercup, St. John's Wort, Ladysmantle, Grass of Parnassus,

Cow Parsnip, Fleabane, Ragwort, Hawksbeard, Sowthistle. She confined her investigations to heights above 6,500 feet: at that level, the vegetation changed rapidly. Above it, on steep slopes, grew thick, rich tropical flora, most of them indigenous. Below, more invasive, temperate species flourished, many imported by American or Australian visitors to the area.

Mary would have been surprised to know that, from our contemporary vantage point, her curiosity could be interpreted as colonial aggression. As the Delhi-born historian Vinay Lal puts it, "The British conquest of India . . . was in every respect a conquest of knowledge, as the land was charted, grammars and dictionaries of Indian languages commissioned, bodies counted, plants and animals (not to mention people) enumerated, and laws codified. The tropes of surveillance, surveying, and statistics appear to encapsulate the chief organizational principles of the modern British state in India, the activities of which were carried out not only by the police . . . and administrators, but by the officials serving under such organizations as the Botanical Survey, the Trigonometrical Survey, and the Geological Survey."[5] This is undoubtedly true; both consciously and unconsciously, Mary played a role in imperial conquest as she strolled among the hills, delighting in color.[6]

On the rare afternoons or early evenings when he could snatch a break from his duties, John pursued the activities that had first charmed Mary when she met him. Years ago, shortly before she accompanied him on the eclipse expedition to Norway, he had published, in *Nature*, a short piece titled "A Remarkable Flight of Birds": "The forms of birds flying at a great height and crossing the solar disk," he wrote, "have been rather frequently seen here [in Kenley, Surrey] during the spring and summer months. . . . They have attracted attention at night, also, crossing the disk of the moon." By studying the shapes of their wings and their manners of flight silhouetted against the sky, John had identified

Fig. 15. The Eversheds' tent in a valley in Kashmir. Courtesy Science Museum Library / Science & Society Picture Library.

"birds of the swallow tribe . . . and others resembling thrush." In his experience, the birds' movements were always "toward the south in August and September," he said.[7] His boyish, amateur's love of *looking* is what delighted Mary most about him.

In Kodai, he noticed how quickly the thunderclouds of April and May seemed to collapse once they had expended their electrical energy. He noticed a surprising scarcity of wasps in Kashmir.

"While walking in the observatory compound my attention was attracted by a cycasnid butterfly of an unfamiliar species, probably a migrant from a much lower elevation," he noted one day. "I was watching the mazy flight of the insect in the expectation that it would settle, when I noticed a shrike sitting on a post nearby, also observing it attentively. He evidently had a fair knowledge of the local butterflies, and considered this something new and worth eating, for he suddenly jumped from his perch and very cleverly caught the butterfly on the wing, a surprising feat for a bird having a rather clumsy build and heavy flight. Apparently he swallowed the insect entire, for I could discover no wings at the spot afterwards."[8]

At night Mary read to take her mind off the distant howling of the jackals or the rustling of the grasses that so often sounded like approaching footsteps. One cold evening, perfect for winter's tales, she became curious about certain references in Shakespeare that suggested he had some knowledge of Copernicus's *De Revolutionibus* (1543), in which Copernicus advanced the radical notion that the Earth orbited the sun.

For example, in *Troilus and Cressida*, when Troilus states, "The bonds of heaven are pec'd, dissolved, and loosed," was it possible that Shakespeare was meditating on the consequences of heliocentric models of the universe?[9] Even if the playwright *had* heard of such theories, he could not have lauded them without fear of official censure. Instead, he would have raised them in code—in poetry.

In *Hamlet*, Laertes, speaking of love, says, "This temple waxes." Then he describes a "crescent."[10] Venus was the star of love. Was it conceivable that Shakespeare knew the phases of Venus—long before Galileo turned his telescope to the sky?

Gently, before turning in for the night, John would warn Mary not to read too late: she'd strain her eyes, he said. Eventually, she set aside the Bard. She'd never be able to gauge his astronomical knowledge—not fully. Her favorite poet, though, was quite another matter.

Dante's excessive reading had seriously weakened *his* eyes. In the *Convivio*, he had written that stars looked as blurry to him as inked letters bleeding into the threads of a damp page. Still, no one had ever studied the world around him more keenly, Mary believed. She returned to her notebook on Dante, and her meditation on the "Quaestio." More than anything, she shared with her poet a passion for *observing*.

If Dante seemed to make a sad mistake about the moon, she figured it was not sloppiness on his part: it was his absorption of the ninth-century Arab astronomer Alfraganus, a major source of Dante's celestial knowledge. It was to Alfraganus that the mistake could really be laid, for he was the one who'd said the moon's perigee was always in the south. In her notebook now, Mary wrote that Ptolemy's lunar circles, as described by Alfraganus, were complicated; but this was no surprise since the moon's motion was, in fact, *very* complicated. She found it remarkable that the irregularities caused by the elliptical form of the moon's orbit had been observed, and a geometrical system invented to represent them, however mistakenly, as early as the second century A.D.

Satisfied that her poet's scientific integrity remained intact—he was only following his most reliable source—she reviewed the other entries she had made in her notebook concerning Dante. As the jackals moaned in the hills, the poet's curiosity, his *companionship*, consoled her. He had once written that he could endure exile from Florence because he could gaze upon the "mirror of the sun and stars and contemplate, under any sky, the sweetest truths."[11] On her hilltop in India, far from home and from other women of her station, she could well understand the exile's dilemmas and delights. She could appreciate the nomad's rage for order—the way he had constructed his poem as a Memory Palace, a Gothic cathedral: *Inferno* with its gargoyles; the pristine ornamentation of *Purgatorio*; the stained-glass dazzle of Paradise, winding ever upward. (It did not escape her that many observatories were also so designed: the writhing cables and howling machinery on the ground floor, untouched by natural light; above this, the telescope platform, scaled to obtain better sight; and finally the great, majestic doors of the dome—when they open, the stars rush in . . .)

Perhaps, Mary thought, closing her notebook, a lengthy study of the *Comedy*'s heavens would tap her astronomical knowledge in a useful way, combine it with her literary sensibility, and provide an excuse to plunge deeper into the pleasures of reading.

A few nights later she mentioned her analysis of the "Quaestio" to her husband. In December 1911, a small article on the "Quaestio" appeared in a magazine called the *Observatory*: "Dante and Medieval Astronomy." The byline read M. A. Evershed and J. Evershed.

Not long afterward, as the observatory's daily routines engulfed John further, he conceded: Dante belonged to his wife. While he snapped pictures of solar flares, Mary settled into a pattern of deciphering Dante, communing across time with him through the shared pleasure of sky-gazing. His "problem[s]" were "much on my mind," she said, as she helped her husband prepare photographic equipment "in beautiful early dawns."[12]

12

Harmonic Structures

For readers of English, the definitive account of Dante's astronomy before 1913 was Edward Moore's investigation of the subject in the third volume of his *Studies in Dante* (1903). Moore acknowledges the complexities that drove T. S. Eliot batty a few years later, saying, "It is a matter of regret that even students of ability and culture often refuse so much as to attempt to understand Dante's astronomical references. They assume . . . that they are not to be understood at all, or at least not without special astronomical or mathematical training."[1]

In his illustrious career, Moore wrote studies of Aristotle's *Ethics* and *Poetics*, and established himself as perhaps the finest nineteenth-century authority on *The Divine Comedy*, editing the comprehensive *Oxford Dante*. Mary knew his work well.

He establishes that, in reading Dante, one accepts the Earth as the cosmos's center. Surrounding it are the crystalline planetary spheres, appearing to move separately from one another against the Fixed Stars. That Dante uses astronomy metaphorically and poetically—sometimes at the expense of observational accuracy—is undeniable. That he means to call the reader's attention to parallels between the spheres' harmonic structures and his own poetic outline is also plain.

Moore emphasizes the *Comedy*'s allegorical beginning: "Midway along the journey of our life / I woke to find myself in some dark woods."[2] As his travels proceed, the Pilgrim emerges as more than just a faceless Everyman. He is Dante, specific to a time and place (time measured astronomically in the poem).

He is named directly—just once—in *Purgatorio* XXX.55, when Beatrice admonishes him not to weep over Virgil's disappearance.[3] Moore suggests—and Mary would agree—that astronomy in the *Comedy* should be read like the Pilgrim's character development: as spiritual allegory underpinned by the educational rigor specific to Dante's era.

Moore spends much of his time deciphering Dante's seasonal measurements. For example, in *Purgatorio* XXXII, Dante speaks of "trees on earth" when the "strong rays fall, mingled with the light / that glows behind the heaven of the fish."[4]

Aries is the constellation glowing behind Pisces, that is, the "heaven of the fish." When the sun is in Aries, its "strong rays" falling, the trees begin to "swell." As Moore explains, "This amounts to saying that the buds begin to swell at the end of March, and the flowers to come out in early April." He concedes that Dante may be "hard," but "he is seldom, if indeed ever, 'obscure.'" No writer "ever had more entirely clear ideas on every subject on which he speaks," Moore says. "In theology, in scholastic philosophy, in metaphysical, moral, and physical science, and in classical literature, if judged by the standard even of a contemporary specialist in each, he will not be found wanting." Dante's knowledge is "extensive," "varied," and "profound."[5]

Though Mary appreciated Moore's work on Dante, she did not slavishly accept his interpretations. She read Dante closely, in English and Italian, and drew her own conclusions. And as we've seen, she was a fastidious recorder of celestial phenomena.

What is it about Dante that inclines readers to take his science seriously? Mary put her finger on his intellectual charm. As she would write in her 1913 volume *Dante and the Early Astronomers*, Dante's description of "celestial matter" is "one of the finest instances of his faithfulness to the teachings of astronomy as he had learned it." In *Paradiso* II, using his "poetical imagination," he examines pearl-like, "polished" ether, a substance "soft as cloud but hard as diamond," which "offers no more resistance to Dante as he enters into it than does water to a ray of light." In passages such as this, Dante uses science and "material facts (as he conceived them) to present an allegory of the deepest religious mysteries." Time and again, if Dante's "premises be granted, the conclusions are correct," Mary says. "As regards sun, moon, and planets," they are often "correct even from the point of view of modern knowledge."[6]

Dante's rigor, and his insistence on specificity even within a dubious framework, seduces readers into casting off what they know, to think in *his* terms.

Like Mary, Dante was most intrigued by the sun—after all, as a Florentine, he had the "true southerner's love" of light: it is the "bringer of warmth and comfort," a sign of "renewed hope and confidence," Mary wrote. Under the sun's rays, "the rose expands, the air is gladdened, mists are dispelled, snow melts, and all things are quickened into life."[7]

Moreover, "Dante seems to have taken some trouble to find out the exact period of the sun's revolution," Mary says. "His text-book, the *Elementa* of Alfraganus, only gives it as 365 ¼ days nearly . . . but Dante wished to be more exact, and somehow contrived to obtain a value which was much nearer the modern estimate of 365 days, 5 hours, 48 minutes, 46 seconds." (His method remains unknown.)[8]

Mary thought Dante must never have seen a solar eclipse, for there is "no allusion [in the *Comedy*] to the features which chiefly strike modern astronomers—the pearly corona, and blood-red prominences standing out like flames round the black disc."[9]

She fastened much of her attention on Dante's Heaven of the Sun. There, he had established the foundation of his cosmos. It was based on mathematics, about which he had written, "The sun . . . illuminates with perceptible light first itself and then all the celestial and elemental bodies. . . . All the other stars are infused by the [sun's] light, and the eye cannot look upon it. . . . [Just so,] all sciences are illuminated [by arithmetic] because all their subjects can be considered under some numerical aspect and in considering them we always proceed by numbers."[10]

Near the beginning of his sun episodes, Dante introduces the topic of geometry, but not before he has asked us to imagine two constellations in the night sky encircling one another, "the rays of one reflected in the other, / . . . both revolving in such a manner / that one went first and the other followed."[11] These systems are one and the same, with segments fast and slow.

Dante explains he is offering this image to make us see the configuration of souls he witnessed in the Sphere of the Sun. It anticipates the final representation of cosmic structure in the poem's last verses.

Following the double constellation, Dante turns to geometry: he insists that mathematical theorems be considered in the context of revolving celestial wheels and the architecture of the universe.

What happens in the Heaven of the Sun? There, the spirit of Thomas Aquinas discusses with Dante the Pilgrim the nature of human intelligence. (In actual life, Aquinas's attempts to harmonize Christian theology with the natural sciences strongly impressed Dante.) Aquinas tells the Pilgrim that King Solomon asked God for wisdom, for the "heart" to judge his people. He did not waste the Lord's time seeking esoteric knowledge such as the "number of angels" in existence, the subject of petty Scholastic debates, nor did he want to know "if in a semicircle a triangle can be formed / without its having one right angle"[12]— a reference to Euclid's *Elements* (300 B.C.). There, Euclid states that all triangles inscribed in a circle using a line bisecting that circle as their base will have, as their apex, a right angle. The point is this: in *not* seeking answers to such problems, Solomon demonstrates wisdom. He rejects Euclidean geometry as a possible source of Truth.

(Just for the record, the task imagined here—drawing a triangle without a right angle within a semi-circle—is impossible in Euclidean geometry.)

Dante, then, knew his Euclid. Apparently, he did not grant Euclid's theorems much credence. So what *did* form his celestial foundations?

Mary is our surest guide through this thicket. In *Dante and the Early Astronomers*, she lays the groundwork for her discussion by explaining that the twelfth and thirteenth centuries, particularly the years just prior to Dante's birth, were periods of turbulent intellectual ferment.

Around 1200, Latin translations of Aristotle's centuries-old treatises on natural philosophy, such as *On the Heavens, Meteorology*, and *On Generation and Corruption*, began to ripple through Western Europe. Aristotle drew on Plato, whose *Timaeus* (c. 360 B.C.), first translated by Cicero (106–43 B.C.), was, prior to Aristotle's writings, the most influential cosmological study available to medieval Christians.

The *Timaeus* proposed a mathematical view of the cosmos. It argued that an intelligent being had created the universe; that the heavens were spherical in structure, aesthetically perfect; and that God had set in place natural laws He would never violate, though He had the option. The *Timaeus*'s insistence on design made Plato's spatial structure wholly compatible with Christian theology. Throughout the early Middle Ages, allegorical interpretations of stellar principles far outweighed literal analyses. For example, in the early fifth century, a Latin translator named Macrobius wrote a popular commentary on Cicero's *Dream of Scipio* (54–51 B.C.), in which a man journeys through heaven, hearing the Music of the Spheres—proof of the universe's mathematical perfection. Macrobius's commentary further advanced Plato's allegorical approach to the stars.

Aristotle's *On the Heavens* (350 B.C.), along with Euclid's *Elements* (c. 300 B.C.) and Ptolemy's *Almagest* (2nd cent. A.D.), introduced greater rigor into Western Europe's cosmological thinking, many hundreds of years after these writings were penned. These works and the works of many other Greek philosophers had been lost to Europe but were preserved in the Islamic world. From about 750 until the tenth century, scores of science and math texts were translated from Greek into Arabic in Baghdad. Starting in the twelfth and thirteenth centuries, with translations by Gerard of Cremona and William of Moerbeke, these treatises, along with those of Alfraganus and other Arabic philosophers, made their way from Arabic into Latin. By 1255, just six years before Dante's birth, the University of Paris had made the teaching of Aristotle compulsory.

Mary carefully sketches all this in her book to provide background for the vision of the heavens that Dante inherited.

Aristotle's universe was largely Platonic but with significant differences. It was spherical yet finite and timeless, composed of some quintessence called ether. Nothingness, vacuums, voids—these concepts had no place in Aristotle's philosophy. He insisted *everything* was solid.

He did not displace Plato's Intelligent Creator but instead posited a Prime Mover, like the still point at the center of a circle around which everything spins. The Prime Mover set the heavens in motion. Fundamentally, then, Aristotle's universe was less allegorical and more mechanical than Plato's.

Ptolemy's *Almagest*, building upon Aristotelian and Platonic foundations, upped the technical ante by cataloging planets and stars (Ptolemy listed 1,022 stars, along with their positions and relative magnitudes).

In these exciting new diagrams of the heavens, it was easy to lose sight of God. Instead of the Holy Trinity, the universe seemed to depend on the trio of Aristotle, Ptolemy, and Euclid. Aristotle's ideas were the most thorough, so he became a particular focus of concern among the pious. A Christian backlash against Aristotelianism was inevitable. The Condemnation of 1277 issued by the bishop of Paris, Étienne Tempier (when Dante was sixteen years old), identified 219 blasphemous propositions. The Church accused Aristotelians of arrogantly *requiring* God to obey natural laws, thereby limiting His scope (his *potentia Dei absoluta*). Some Aristotelians even suggested that, since no evidence of other worlds existed, God could not create them. Article 14 of the Condemnation made it mandatory to profess that God could create as many worlds as He pleased. Just because He *had* not done a thing did not mean that He *could* not do it.

In effect, and inadvertently, Article 14 loosed the human imagination—it encouraged speculations about possible realities and alternatives to the known universe, all of them viable should God choose to pursue them.

To a subtle mind like Dante's, the Condemnation did not cancel Aristotle's universe—it expanded it, allowing the addition or evolution of fanciful new notions. It gave him, in effect, carte blanche to refine existence for his own poetic needs.

In Dante's model, the Earth stood still at the center of the material universe, circled by seven planetary spheres—those of the moon, Mercury, Venus, the sun, Mars, Jupiter, Saturn—and an eighth sphere, that of the Fixed Stars. To this Aristotelian-Ptolemaic concept, medieval Christians (including Dante) added a hierarchy of angels. The angels moved the individual spheres, while the different planets possessed unique qualities exerting influence over the Earth.

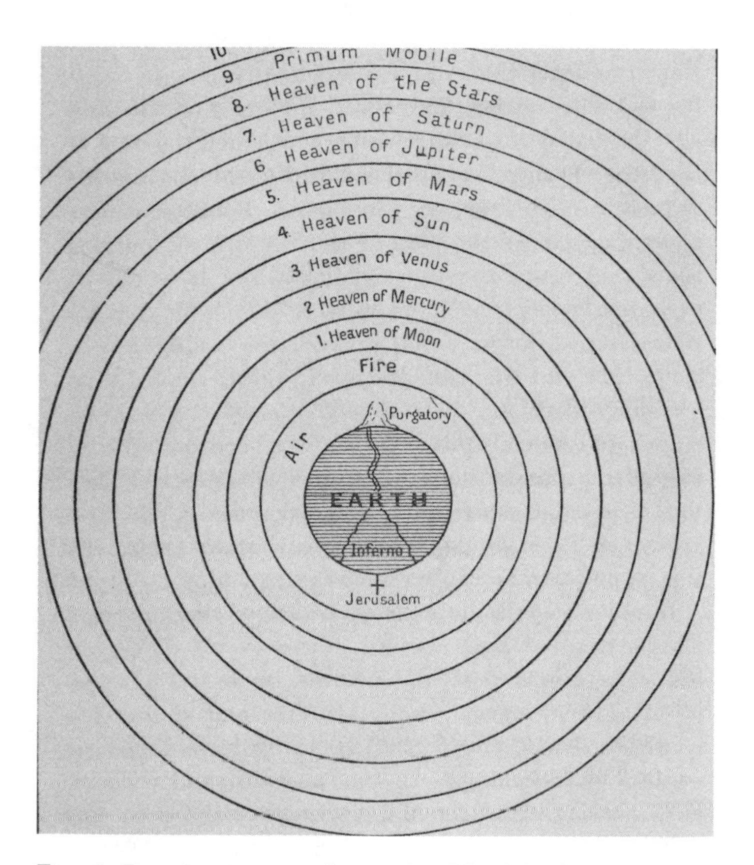

Fig. 16. Dante's universe as illustrated in Mary's book *Dante and the Early Astronomers.*

In *The Divine Comedy*, Dante added a ninth heaven, the Primum Mobile, a crystalline sphere, luminous and transparent, that moves the angels. Mistakenly, in the *Convivio* (2.3.3–6) Dante attributed the idea of the Primum Mobile to Ptolemy. In fact, Arab astronomers created it initially to address the problem of precession, when planets and stars appear to revolve from east to west in a twenty-four-hour period (a result of Earth's rotation). Mary says Dante's mistake suggests that he never read

Ptolemy directly, but knew him through secondhand sources, especially Alfraganus's *Elementa Astronomica*, translated into Latin by Gerard of Cremona (c. 1114 – 1187).

It's when Dante tries to describe the Empyrean, the abode of God and the angels beyond the Primum Mobile, that matters take a real twist.

Or—a curve.

Medieval Christians had established a tenth sphere, the heaven of God's perfect light, where angels, saints, and the righteous dead dwelled forever. The problem was, if this realm lay beyond the Primum Mobile, just where was it? Aristotle had said the universe was finite. Most scholars agreed (though this caused a new fix—where and what was the edge of the universe? what happened there?). If the Empyrean wasn't nestled within the visible heavenly spheres, did it float in a void—in nothingness? As we have seen, Aristotle rejected such notions, believing that the universe was finite and all its contents solid. But if the universe *was* finite, where was its place? How was *place* defined? Could the universe be both finite and unbounded at once? This seemed impossible.

It's when Dante reaches the apparent edge of *everything* and begins his approach to God that he becomes most radical, conceiving a cosmos never seen before.

To fully appreciate Dante's vision, we must leap forward in time for a moment, past Mary on her hilltop, to our own day. We need to adopt the equivalent of the angelic consciousness, keeping in mind past, present, and future.

In 1992, astronomers released a sky map of the Cosmic Microwave Background. Informally, they referred to this as the "afterglow of the Big Bang," from the "time that light and matter parted ways."[13] George Smoot, a Nobel Laureate who led the research team, said, "If you're religious, this is like looking at God."[14]

Discovered in 1965 by the American radio astronomers Arno Penzias and Robert Wilson, the Cosmic Microwave Background is the earliest radiation that cosmologists have detected, left over from the Big Bang, and it fills the entire universe. According to the Big Bang theory, atoms could not exist under the intense temperatures and pressures characterizing the first 300,000 years of the universe. Instead, matter was distributed as ionized plasma—that is, a highly charged gas, featuring strong electromagnetic dynamics that scattered radiation. As the universe expanded, temperatures and densities dropped, so electrons and atomic nuclei were able to form atoms. At that point, photons—essentially, carriers of electromagnetic force—were able to escape what cosmologists tend to call the "fog of the early universe" and travel about, unfettered. The Cosmic Microwave Background is the record of the photons' moment of escape.[15]

It reveals, then, a sliver of the universe as it looked 13.7 billion years ago, which astronomers calculate was roughly 400,000 years after the Big Bang. At the moment of the Big Bang—what we might call the beginning of time ("Events before [that] are simply not defined," writes Stephen Hawking)[16]—the universe as we know it sprang into being. It has been expanding ever since. The farthest galaxies we are able to detect are remnants of the universe's earliest moment. One is tempted to say they approach the cosmos's birth point, but this would be misleading since the Big Bang happened everywhere. Like liquid scatter after a splash, they are the outermost droplets tossed from the violent eruption. Because they're so distant, "they span our entire field of view, in all directions of the sky," writes the science reporter David Castelvecchi. We perceive ourselves, as observers, at a central point; from where we perch, these remote galaxies occupy our circumference.[17]

Here is the paradox: although they fill our entire span of vision, they are really a very narrow slice of space-time—the piece that existed just after the precipitating maelstrom. "As we look out at the universe, we are looking back in time, because light had to leave distant objects a long time ago, to reach us at the present time," Hawking writes. "As one goes back into the past, the area of our past light cone will . . . start to decrease."[18] From our limited post, at what seems to us a fixed point (though of course it is not really), the larger our field of apprehension, the narrower our view. Our circumference appears to be our center. The outer "sphere" of our perception is bounded by its interior.

Fig. 17. Gustave Doré's illustration of *Paradiso* XXXI, Dante and Beatrice crossing the Primum Mobile into the Empyrean.

Thus, the rules of Euclidean geometry shiver, shatter, disappear —a possibility that Dante alerted us to centuries ago in his Heaven of the Sun.

In *Paradiso*, after Dante the Pilgrim crosses the Primum Mobile into the Empyrean, the spiritual world where God, the angels, and the virtuous dead reside, he observes that this other realm, like the material universe, has a geometric structure, a mirror image of the physical world, with nine concentric spheres. But instead of growing ever larger, as Euclidean geometry tells us they must, they grow smaller. "And at the center, says Dante, sits God, occupying a single point and emanating a blinding light"—eerily similar to the Big Bang, Castelvecchi writes.[19]

In the early twentieth century, Einstein's theories of relativity would sketch a non-Euclidean universe, one in which space is curved, allowing it to be both finite and bounded.

Dante signals *his* deep interest in curved space in Book II of the *Convivio*, written about fifteen years before the completion of *The Divine Comedy*: "Geometry moves between two things antithetical to it, the point and the circle: as Euclid says, the point is the primary element in Geometry, and, as he also indicates, the circle is its most perfect figure, and must therefore be considered its end; and both of these are antithetical to the certainty characteristic of this science, for the point cannot be measured at all, since it cannot be divided, and the circle cannot be measured precisely, since, being curved, it cannot be perfectly squared."[20]

This passage is decidedly odd. First, Euclid never wrote the things that Dante claims he did. Dante misuses the geometer to advance his own ideas. Second, *paradox* clearly fascinates our poet. The science of geometry is characterized by certainty, he says, yet its methods fail. In Dante's language, beginning and end (point and circle) are related, perhaps the same.

In *Paradiso* XXVIII, Dante, in the Primum Mobile, gazes behind him at the "little threshing floor" of the Earth. Then he scans upward toward the bright point of God. Yet in looking *up* he is also looking *back*, like someone who sees a candle flame in a mirror and "turns to see if the glass is telling him the truth."[21]

In curved space, direction blurs.

In the Primum Mobile, at the apparent edge of an edgeless universe, Dante occupies no particular location. He puts it this way: "[I was] thrust into heaven's swiftest sphere [the Primum Mobile]. / Its most rapid and its most exalted parts / are so alike I cannot tell / which of these Beatrice chose to set me in."[22] Every point is the same in a finite yet unbounded universe—curved space means that if you travel as far as you can from where you began, you will wind up right where you started.

Beatrice tells Dante, "The nature of the universe, which holds / the center still and moves all else around it / starts here as if from its boundary line."[23] The middle is linked to the outer limit. "Dante has invented a new topological space, the 3-sphere [or the hypersphere]," says physicist and mathematician Mark A. Peterson. "If his imaginative feat had been recognized in his own time, and if the idea had been pursued and developed, Dante would today be considered one of the inventors of topology, and one of the great creative mathematicians of all time. As it is, he is not even a footnote to topology, which was only invented officially in the eighteenth century, and didn't really take off until the twentieth," tucked into Einstein's thoughts.[24]

In *Planets, Stars, and Orbs: The Medieval Cosmos, 1200–1687*, the science historian Edward Grant writes, "When, in 1632, Galileo published his *Dialogue on the Two Chief World Systems*, his monumental assault on Aristotelian cosmology, the majority of natural philosophers, if not the majority of astronomers, were probably defenders of some form of geometric cosmology."

Gradually, over the next fifty-five years, Aristotelian concepts eroded, until in 1687 "Isaac Newton's ... *The Mathematical Principles of Natural Philosophy* ... [gave us] an ideal terminal date" for Aristotle's views and full acceptance of Copernicus's sun-centered system.[25]

But with the warehousing of medieval cosmology, many useful ways of knowing vanished. How else to account for the fact that, six hundred years after Dante finished *Paradiso*, his science came to seem relevant again—not outmoded at all, but remarkably contemporary, even prescient?

From the seventeenth century on, the dominant opinion was that "medieval thought had little if any importance to the progress of science, and in fact may have been detrimental," writes William Egginton, a humanities historian. Yet Albert Einstein's relativity "posited space and time as observant-dependent entities. ... This conception and the space of Dante's natural philosophy have far more in common than does the space of the intervening five hundred years, which is why Dante could imagine his cosmos, while lacking the language and tools to call it such, as a hypersphere."[26]

Just as Dante, imaginatively freed by the Condemnation of 1277, accepted Aristotle's insistence that all things, including voids, must be made of matter, and hankered to solve the problem of an unbounded, finite universe, Einstein chafed to push beyond the concepts of *infinity* and *empty space* that dominated modern cosmology. (These ideas had become ubiquitous as a way of dismissing the irritating question of whether or not the universe had an edge.) Like Aristotle, Einstein wanted to show that space-time was a *physical* reality. Like Dante, he embraced the possibility of non-Euclidean geometry. Sitting among her wildflowers in India, writing about Dante, Mary would soon hear all about Albert Einstein: a fellow medievalist, a centuries-old soul.

13

"Dante and the Early Astronomers"

The world's nasty politics whispered in Mary's ear as she readied *Dante and the Early Astronomers*. During the next several years, the events goading nations to war would touch her personally, despite her isolation in Kodaikanal. Paper shortages restricted scientific publications. Amateur skywatchers reported that, in conflict zones, "the difficulties in the way of meteor observation are [becoming] insurmountable: 'there [is] no adequate means of distinguishing between a fireball and a star shell.'"[1]

Frequently, Mary turned to the lovely passage in *The Divine Comedy* in which heavenly movements stir a breeze—a motion, Dante said, that "strikes the thick forest and makes it murmur."[2] She imagined the cooling draught on her face. She ignored war thoughts. The "remoteness and loftiness" of her situation, "surely unprecedented in Dante studies," mused Barbara Reynolds in her introduction to *Dante and the Early Astronomers,* gave the book its "remarkable perspective of vision and serenity of tone."[3]

While John complained about breakdowns of the old Shelton clock in the spectroheliography room—how was he supposed to work with equipment well over 130 years old?—Mary delighted in rereading Dante's description of saints circling the sun like the gears of a giant timepiece (probably the first description of an escapement clock in all of Western literature; Dante saw one in Milan in 1310, when he arrived for the coronation of Henry VII). While John groused about heat-wave damage to his telescope mirrors, Mary reflected that when Dante speaks of mirrors he probably means glass treated with an alloy of antimony, lead, or tin. (For Dante, water in the eyeball was like the glass in a mirror, and the retina like the metallic backing.) While John pressed his eye to the telescope's eyepiece, Mary recalled that, for her poet, optics was solely concerned with desire. It was the nature of light to be aggressive, activating the color inherent in the surface of bodies. Seeing was a willful act, potent, a matter of visual fire shooting from the eye, seizing the object of its delight.

Daily, John recorded sunspot movements. Mary returned to the *Georgics*, in which Dante's beloved Virgil wrote of the sun's "flecked" disc.[4] She urged her imagined readers to appreciate the "changing pageant of the sky": a few direct observations would do more "than any written explanation" to help readers experience Dante's Heaven, she wrote.[5]

Unlike most writers, who "remember imperfectly" what they see of the sky, "or draw upon their imagination without any knowledge of celestial movements," Dante takes great care to avoid incongruities, Mary says. His Beatrice speaks with stunning "scientific precision." For most writers, the "moon is especially a stumbling block, and it is quite a rare thing for a modern novelist to introduce one without making it do something impossible. A new moon will rise at midnight, or a waning moon at sunset." Dante's moon "does indeed give us a little trouble once or twice, but he never makes flagrant mistakes of this kind."[6]

He is not a singer of lunar light, for it "means a loss of starlight, which he dearly loved."[7] Mary could be speaking of herself. As she studied and wrote, her kinship with Dante grew.

She was particularly pleased with his southern skies: accurately, she writes, "he does not place any single bright star to mark the south pole."[8] In *Purgatorio* XXIV he mentions four bright southern stars "non viste mai fuor ch'a la prima gente"—seen by no one except the first people. Scholars have long debated the identity of these stars, and many readers have concluded that Dante is fictionalizing here. Mary says, "It is a fact that there are four bright stars in the form of a cross, lying between 56 degrees and 63 degrees south." Although they "were not recognized as a separate constellation until the beginning of the sixteenth century, when Amerigo Vespucci described them in his letters," she speculates that Dante saw or heard about these stars from travelers. She provides a map "showing what stars would have been visible to Dante at the supposed latitude of Purgatory, when Pisces was on the eastern horizon, as described in *Purgatorio* 1."[9]

From time to time, Mary's precise modern methods catch Dante in "little inaccuracies," but overall she finds him remarkable in his "consistency and truth of description."[10]

Reluctantly, she concludes that the astronomical alignments in the poem do not match the real look of the sky in 1300, the year of Dante the Pilgrim's journey; for example, in 1300 the moon was not full on the night between Thursday and Good Friday, as Dante implies. He is off by a whopping three days! (She says he may have consulted a flawed almanac; or perhaps, more deliberately, he wanted to emphasize the contradiction between the Roman calendar, used by the Pope, and the Florentine calendar, whose new year coincided with Passiontide in March.) Mary argues that Dante is intent on showing how *all* dates, calendars, and means of marking time are highly contingent, based on our imperfect understanding of heavenly movements.

Ultimately, the "heavens are the same to our eyes as to Dante's," Mary says. Although the "ideas suggested" by celestial bodies are "widely different" to us than they were to the poet, our universe is still "vague, vast, mysterious, without known limits or centre, offering problem after problem to the thinker."

Therefore, we "share the feeling of our medieval forefathers, of the ancient Greeks, of the earliest men whenever and wherever they became men. The unerring courses of the stars speak to us . . . of perfect harmony."[11]

The sun seemed to settle down, some, while Mary wrote her book: John noted that "not a single northern spot has been seen" on the sun's surface "from January to November 1912" and solar prominences were few in 1913.[12] If, during this period, the observatory's plates had "traced beautiful fountain jets" of fire, Mary might have had difficulty concentrating on her manuscript, but the heavens cooperated with her project.

Twice, John left her alone to set up camp in the Valley of Kashmir. He found the solar definition extremely good there, eight thousand feet above sea level where locals cultivated rice. At home, Mary read, wrote, drew maps based on Dante's details, and corresponded with scientists and prominent Dante scholars, including G. V. Schiaparelli, a well-known medievalist. Although astronomy was the focus of her study, her ultimate goal was to celebrate Dante's poetry. "No one will dispute a poet's right to arrange the skies as he thinks fit," she wrote.[13] She put the observatory staff to work, checking her interpretations of the *Comedy* against theirs. It is surely an anomaly in Dante studies that names such as Mr. Sitarama Aiyar and Mr. Krishnaswamy appear among the acknowledgments. Her sister Lucy arrived to help with the indexing, and Mary dedicated the book to her.

With her task nearly completed, Mary could no longer ignore the rumors of war. "Dante's world was easier to think about than ours," she wrote. "Encompassed by an infinity which gave the imagination free scope[,] the material universe within was neatly rounded off, as it were, completely finished." She longed for a "harmony" as "grand" as Dante's vision, but that would involve casting away Earth's distressing details.[14]

Gall & Inglis published *Dante and the Early Astronomers* in 1913. As the historian Mary Brück says, "The circumstances of its publication were . . . unfavourable: the author was back in India, cut off by the First World War." Students of literature "were not likely to come across the catalogue of a scientific publisher. It was to be over thirty years before the book came to the notice of Dante scholars." Only two reviews of the book appeared in the science press: a note in the *Observatory* calling it "charming," and a full-page consideration in *Nature*, lauding Mary's valuable contribution to historians of science.[15]

14
Sun-Chasers

After a short trip with John to New Zealand to help the government choose a site for a research station, Mary resumed assisting her husband around the observatory. Using his photographs, she classified solar prominences into active and eruptive types, tracked sunspot motions, and recorded more than sixty thousand prominence observations over an eleven-year sunspot cycle. Her work at Kodaikanal, says Mary Brück, was a "precursor" of "cine-photography with the coronograph, a technique then far in the future."[1]

In the latter months of 1913, Mary accompanied John to Kashmir. He had begun to wonder if observing conditions were better in the valley (also, he had grown tired of leaking roofs). "A doonga was hired, and all instruments and stores put on board, and . . . we started on a prospecting tour up the River Jhelum as far as Islamabad, observing the sun from the river bank at various localities," John wrote. "A very convenient site for a temporary Observatory was found about 10 miles out from Srinagar near the village of Pampur. This was a small grass-covered hillock about 100 yards from the river bank, rising some 20 feet above the general level of the plain."[2] From a short distance at night, with lanterns glowing from within, their tent resembled a low-slung lighthouse casting a pale glare out over the river.

In her pith helmet and long-sleeved satin dress, crawling at dawn through the tent's canvas flaps, Mary knew she looked less like a modern-day astronomer than a figure from Galileo's day. She wondered, as she had many times, if science had really

Fig. 18. Mary with telescope and tent in Kashmir. Courtesy Terry Evershed.

progressed much beyond the Florentine's vain insistence to churchmen that "Aristotle . . . [would] have abandoned" the notion of the "inalterability of the heavens" had he observed sunspots.[3] How far had we come since William Herschel's speculations that there might be cities on the sun?

How "modern" could Mary and John be, she wondered, standing ankle deep in a field of wet rice, brushing wasp nests from their spectroheliograph (though John noticed a continuing "despondency of the wasps" in Kashmir). Mostly, John was obsessed with the proper place to pitch camp; he suspected uninterrupted water surfaces contributed to "good seeing" because they generated an "absence of disturbances in the lower strata of the air." Maybe islands surrounded by calm seas were better spots to observe the sun than mountaintop observatories. Mary suggested they look for Purgatory.[4]

They marched through valleys, plains, and foothills, waded through muddy rivers, attempting to satisfy John's restlessness. His concern with the perfect perch from which to chase the sun would grow more acute the moment he heard of Albert Einstein.

Fig. 19. Mary at a makeshift desk at a Kashmir campsite. Courtesy Science Museum Library / Science & Society Picture Library.

According to reports received from the British Astronomical Association back in London, this obscure German patent clerk had "concluded that a ray of light would be deflected when passing through the gravitational field of the sun" (as if beamed through a lens). He "put forward the value of 0.″87 as the amount of deflection," later refining this to 1.″75. Additionally, Einstein had predicted that wavelengths of light emitted by a massive body would increase by an amount proportional to the local gravitational field. He called this the "gravitational redshift."[5]

From 1915 on, John's energies would be devoted almost exclusively to proving or disproving the "Einstein Effect."

In 1915, Mahatma Gandhi returned to India after spending over twenty years in South Africa, practicing on the white minority government there the nonviolent protest strategies he would employ to confound India's rulers—*Satyagraha*, the Strength of Truth.

It is doubtful the Eversheds or their Kodaikanal crew, reading journalistic accounts of the man, would have thought much of him: he opposed Western scientific techniques, Western education in general, and he seemed leery of progress. His renunciation of physical pleasures was incomprehensible to most Indians, let alone Westerners.

His return coincided with the Raj's decision to pass legislation extending into peacetime the civil restrictions of the war period.

Even in the relatively safe cocoon of the observatory, John and Mary heard more and more accounts of national unrest. "Growing Indian nationalism inspired by new campaigns of civil disobedience could not be ignored even in the hills," writes geographer Judith T. Kenny.[6] In 1910, in the princely state of Bastar, eastern India, the tribal population had risen against a small British military force stationed in the region, in protest of the land displacement caused by English control of the forests. Local journalists referred to the uprising as the Bhumkal rebellion, an earthquake that would, in years to come, lead to more violence.[7]

Meanwhile, in Tuticorin on the Bay of Bengal, a place best known for excellent pearl fishing, a lawyer named Chidambaran Pilan had led a boycott of British goods to promote local produce, or *swadesh*. He founded the Swadeshi Steam Navigation Company to challenge the British trade monopoly.

At the time, the British Sub-Collector in Tuticorin was a prideful man named Robert W. D. E. Ashe. To crush the Swadeshi movement, he ordered authorities to fire into a crowd of striking workers. He also had Pilan arrested on charges of sedition and seized Swadeshi steamships. The government immediately promoted him to Collector and District Magistrate of Tinnevelly (modern-day Tirunelveli, in Tamil Nadu).

One summer day Ashe was traveling by train in a first-class compartment. The train stopped at the railway junction of Maniyachi. A young man rushed into his car and shot him dead at point-blank range. The young man, Vanchinatha Iyer, ran down the graveled platform, chased by police, and into a public lavatory. There, he shot himself in the mouth. On his body was a note: "Every Indian is at the present time endeavoring to drive out the Englishman who is the enemy of [our] country and to establish 'Dharma' or liberty."[8]

Throughout Tamil Nadu, in the hills around Kodaikanal, police searched caves, ravines, villages, and even temporary scientific outposts for social extremists and conspirators. John's sun-camps, not only in Kashmir but also closer to Kodai, came under such intense scrutiny he lost his appetite for them. Besides, by now he had concluded that the "four months [from] December to March ... must be considered less favorable in Kashmir than at Kodaikanal because of the greater prevalence of clouds in Kashmir at that season."[9] Still avid to pursue the "Einstein Effect," he restricted his experiments to—and protected his wife's safety on—his remote mountaintop.

15

Exploding the Sun

When Galileo was old and blind, languishing under political arrest in the tiny suburb of Arcetri, near Florence, he received a brief visit from the twenty-nine-year-old English poet John Milton. Afterward, Milton wrote that the Catholic Church and its Office of the Inquisition had hobbled the "glory of Italian wits" by silencing the great scientist.[1]

Milton understood that Galileo had recreated the universe. He learned from the old man that Dante had been an inspiration to him. Later, when Milton revised Heaven in his grand poem, he leaned on Dante, too.

In passages on the Heaven of the Sun, Dante had written, "In the court of Heaven . . . / many gems are found of . . . worth and beauty."[2]

Dante does not mention sunspots, but in *Paradise Lost*, Milton, mindful of the poet and the scientist he had inspired, related sunspots to mosaiclike gems—they are luminous and "beyond expression bright."[3]

Galileo attempted many experiments in his lifetime, and some of them failed. One major failure was his proposal to measure the velocity of light. He wanted to know if light's propagation was instantaneous.

Here was the test: He placed two men at a certain distance from one another. Each held a light he could easily cover and uncover. On command, one of the men would unveil his light; the instant the second fellow saw it, he would follow suit. After that, Galileo increased the distance between the men. Would the greater separation cause a longer delay between light signals?[4]

The relatively short distances involved and the subjects' imperfect reaction times guaranteed that Galileo's experiment would sputter. But his methodology was logical. In 1675, the Danish astronomer Ole Roemer applied Galileo's approach to interstellar "lamps"—that is, to Earth and to one of Jupiter's moons, blocked and unblocked by eclipses.[5] Today we do know the speed at which light travels—186,000 miles per second. Galileo would have been pleased to know such a specific figure. Einstein would show that nothing moves faster.

In the early twentieth century, Einstein also discovered that light has a conflicted nature, sometimes behaving as a particle—individual chips, as it were, adhering to create the appearance of a wave or a stream, like Ravenna's heavenly mosaics. His findings indicated that light is both a particle *and* a wave—and neither.

Many years prior to this discovery, in 1801, a British physicist named Thomas Young had conducted what he called the "double slit experiment."[6] Young believed that if light were wavelike, it should behave like water ripples. When two ripples meet from opposite directions, they either reinforce or destroy each other. Directing sunlight through a series of slits, and then projecting the light onto screens, Young concluded that his results illustrated light's wavelike qualities. Nearly a century later, Einstein found that when ultraviolet light shines on a metallic surface, electrons can be ejected from the metal's face—the photoelectric effect—and this suggested that light acted much more like a series of particles.

At the time of his discovery, Einstein was a menial worker in Bern, Switzerland. A proud man, every bit as haughty as Dante and Galileo (surely, all three men are doing diligent penance on Purgatory's first terrace!), he had irritated many people in the world of science. He couldn't get a university job. His minor

Fig. 20. Einstein in his office at the University of Berlin, 1920. From *Science Monthly*.

office post gave him plenty of time to contemplate theoretical models.

He began to focus on light. In particular, he thought about the sun.

In the Sphere of the Sun, Dante plants his little explosion, that is, his repudiation of Euclidean geometry. In a similar thought-experiment, Einstein exploded our mother-star in a heroic effort to fully grasp light, space, time, and the workings of the universe.

(We don't know if Einstein ever read Dante seriously, though we know his sister, Maja, studied *The Divine Comedy* with physicist Wolfgang Pauli's wife, Franca. Einstein's reading tastes ran from fatalism to nihilism: Goethe, Dostoevsky, Nietzsche.)[7]

What would happen to us, here on Earth, if the sun disappeared in the flick of an eyelid, Einstein wondered. Would we know it right away? The sun's distance from Earth, approximately 93 million miles, means its light beams take about eight minutes to reach us. If the sun vanished, is it true that we wouldn't know it until eight minutes later, since nothing travels faster than light? Or, without the sun pulling on us, would we suddenly careen out of normal orbit for eight minutes before we witnessed, via light, the sun's demise? Common sense, not to say Newton's laws of gravity, suggested that a solar catastrophe would launch Earth like a rock from a sling. (And if something moved faster than light, it would violate the laws of cause and effect.)

This puzzle preoccupied Einstein for years.

Finally, in 1915, under deadline pressure to complete a lecture, he worked through the problem using Galileo's passion for observation and Aristotle's belief that *matter* mattered. (His failure, late in life, to embrace the implications of quantum mechanics came from his kinship with Aristotle, Aquinas, and, ironically, the Catholic Inquisition, all of whom wanted the

Fig. 21. Space-time curvature. Courtesy NASA.

universe to be perfect, operating under smooth, predictable laws.)[8]

Still thinking about the components of light, Einstein concluded that space is not flat or passive or empty. Rather, it is curved. Dynamic. It tells matter how to move. Imagine a sheet of fabric. Imagine placing a steel ball on the fabric, causing the sheet to bend. The presence of a planet or a star or any other heavy mass affects space the same way. Earth circles the space-warp made by the sun like a marble rides a groove.

In such a universe, parallel lines, like rays of light, can meet; the sum of the angles in a triangle can amount to more or less than 180 degrees, depending on *how* space flexes. As Dante had hinted over six hundred years earlier, existence is non-Euclidean.

Time is motion, Einstein said. We sense time passing because objects change their positions. And motion is susceptible to the distortions of ripples in space.

One major implication of these concepts, writes Nigel Calder, paraphrasing Einstein's discovery, is that "time would stop if you could travel at precisely the speed of light. You would abolish distance entirely, so that your point of departure and your destruction seem to be at the same place."[9]

If Einstein's general theory of relativity was right, light would bend around any massive object. The sun was certainly massive enough to divert starlight's path. If there was a way to shade the brilliance of the sun itself, Einstein said, we would see the constellations move.

John Evershed wasn't so sure.

16

Saturnalia

Even in the best of times, getting from Kodaikanal back to Britain was arduous. Motor cars and buses were rare sights on the roads of southern India. John and Mary traveled by bullock-carts, drawn by oxen, to the railway station at Periyakulam, several miles southeast of Kodai and several thousand feet lower in elevation. Pack ponies carried their baggage. Coolie teams would have to switch out at the journey's halfway point, where the road zigzagged through steep rock cliffs and no trees rose to shade the sun. As evening approached, the coolies would light torch flares. Rumors of dacoits—armed robbers—caused Mary no small anxiety. She was always immensely grateful to reach the station's cool retiring rooms.

By 1915, the smoldering effects of war in the cities of Europe had made international travel even more difficult.

John and Mary tried each spring to get to Sion College in London for meetings of the British Astronomical Association. Einstein's theories now dominated talk at the meetings. The association's president admitted, "An astronomical matter . . . [is] engaging all [our] attention, but one which [we] are ashamed to say [is] beyond the limits of our understanding."[1]

London had been pounded, during the war, by German Zeppelins ("baby-killers," the British papers called them) and by Schütte-Lanz airships. Eventually, more than five thousand bombs would be dropped on English towns. Incendiaries sparked fires in London suburbs. Families feared they'd burn forever. In a letter in September 1915, D. H. Lawrence described a raid to a friend: "We saw the Zeppelin above us, just ahead, amid a gleaming of clouds, high up, like a bright golden finger. . . . Then there [were] flashes near the ground—and the shaking noise. It was like Milton . . . war in heaven. . . . I cannot get over it, that the moon is not Queen of the sky by night, and the stars the lesser lights. It seems the Zeppelin is in the zenith of the night, golden like a moon, having taken control of the sky; and the bursting shells are the lesser lights."[2]

That same year, the director of the BAA's Saturn Section (a subgroup devoted to studying the ringed object) "was called away on military duties," the association president informed his colleagues.[3] The Eversheds canceled their annual journey to England. It was as if Saturn, the pale, melancholy planet, had grazed Earth's atmosphere, filling it with poison. A yellow malaise enveloped the face of the waters.

At Kodaikanal, John lost his assistant director, T. Royds, to army service in 1917. And then, in October 1918, influenza claimed his second assistant, G. Nagaraga Aiyer. Aiyer had worked at Kodai since 1899, before the observatory was fully operational. It is impossible to read John's annual reports from this time and not hear, beneath the dry tone and passive voice, his personal anguish. Public Works "left [our leaking roofs] in an unfinished and very unsightly condition," he writes. "The wire fence around the compound is sagging—Public Works did not respond to requests to fix it."[4]

It wasn't just the fence that sagged.

John was obliged to report that the "head peon who also acted as engine and dynamo attendant died . . . from pneumonia. The accommodation for such cases at the Kodaikanal Municipal Hospital is quite inadequate, and it is considered that the life of this valuable and efficient servant might have been saved with reasonably up-to-date arrangements and nursing."[5]

His complaints went on and on. He could barely suppress his frustration and despair at even the smallest problems: "There has been great delay in installing a new pump by the Public Works Department, and much difficulty is experienced in carting water for photographic purposes."

Instead of concentrating on Einstein, he was, literally and figuratively, smothering grass fires.

In Berlin, where Einstein had taken a research position near the beginning of the war, the physicist faced obstacles of his own. His theories were being attacked on racial grounds even by some of his fellow scientists. A prominent German anti-Semite named Philipp Lenard, who had won a Nobel Prize for his work on cathode rays, dismissed Einstein's "Jewish way of doing science"—that is, "spinning webs of abstract theory that lacked any roots in the firm and fertile soil of experimental work."[6] In time, these hateful professional critiques would devolve into threats on Einstein's life. Lenard hoped the defeat of the British in the war would serve them right for never—in his view—sufficiently recognizing his own work.

For their part, the few British scientists who knew about Einstein's general theory of relativity refused to consider its consequences because they were furious about the German Zeppelin raids.

17

Infinity and the Fly

Before 1919, general relativity was not well known in England. The war had prevented German scientific journals from reaching London. Willem de Sitter, a Dutch astronomer, obtained Einstein's 1905 relativity paper and sent it to Arthur Stanley Eddington, secretary of the Royal Astronomical Society, around 1916. At the time, this was Britain's only hard copy of the universe slouching to be born.

Einstein reiterated that, to see the sun bend light, observers had to shield the sun's brilliance. He suggested that his hypothesis might be tested during solar eclipses, as the moon passes in front of our nearby star. In the summer of 1914, in fact, a party from the Berlin Observatory had traveled to southern Russia to observe the total eclipse on Einstein's behalf. When Archduke Ferdinand was assassinated on July 28, Austria-Hungary invaded Serbia; on August 1, Germany declared war on Russia. The astronomers were interred, their equipment impounded.

Later, learning of these developments, Arthur Eddington was appalled. A member of the Religious Society of Friends, better known as the Quakers, he believed all people possessed divine potential, an actual inner light. He was also staunchly opposed to war. It upset him to see British astronomers publish anti-German diatribes in the pages of the *Observatory*—for example, "Is it not a fact that babies have been killed in ways almost inconceivably brutal ... as a part of the deliberate and declared policy of the German army? ... Is it not a fact that German men of science have gone out of their way to declare their adhesion to these things?" Eddington responded to such comments by penning letters to the *Observatory* urging international scientific cooperation as the way to counter war fever. "Surely Professor Eddington is here using . . . his own shrinking from horrors to help him in ignoring hard facts [about the Germans]?" one of his colleagues replied.[1]

Eddington declared himself a conscientious objector. Scorn ensued from many Royal Society men. Nevertheless, he had a strong supporter in the Astronomer Royal, Sir Frank Dyson, who considered him his scientific protégé. Dyson exercised considerable effort with the British government to keep Eddington out of jail following his refusal to serve in the army, promising officials that, if allowed, Eddington would conduct a crucial scientific experiment of world import.

By this time, with the approach of the 1919 solar eclipse, Eddington had immersed himself in the general theory of relativity. He found it compelling; more than that, he believed if a British research team could prove a German physicist correct, science might ease the world's tensions. He admitted he was biased in favor of Einstein's theories: they offered a prime opportunity for peaceful cross-cultural relations.

One of the men fouling Eddington's agenda was John Evershed, sitting on his hill in India. In 1911, when Einstein had predicted a solar redshift, John turned his mind to the problem. Initial results with the spectroheliograph corresponded closely to Einstein's values, but there were enough anomalies within the experiments to cast doubts on John's accuracy. He was frustrated by the limitations of his equipment. His efforts prompted Charles St. John at Mount Wilson in Los Angeles, working with a state-of-the-art sixty-foot tower telescope and a high-dispersion spectrograph, to tackle the redshift. His results radically opposed Einstein's numbers. Eddington worried ("this is a rather severe blow to those of us who are attracted by the relativity theory," he wrote a friend). He kept a close eye on Mount Wilson and on John's continuing efforts at Kodaikanal.[2]

John irritated him also by suggesting, in the pages of the *Observatory*, that his intended locations for observing the approaching eclipse were not optimal. The best place to view a solar eclipse, John said (based on his experiences in Kashmir), was "at a sea-level site near a body of water, when daytime temperatures would be most consistent"—for instance, near a rice field "flooded so that the valley becomes a virtual lake." This way, heat waves would be small, lessening the atmospheric turbulence.[3] In spite of Eddington's international goodwill, he didn't like this bloke off in the colonies presuming to tell him how to conduct his business.

On March 8, 1919, Eddington set sail from Liverpool aboard the HMS *Anselm*, headed for the Portuguese colony of Principe, in the Gulf of Guinea off the west coast of Africa.[4] He was leaving to avoid prison, to fulfill his social obligation, and to remake the universe in Einstein's image. The core of the experiment was to test Einstein's prediction that starlight traveling to Earth would be deflected a certain amount by the sun's gravitational field. The sun's blackened disc would allow stars near the edge or limb (in this case, the Hyades star cluster) to be measured.

Eddington carried with him a single instrument, an astrographic telescope borrowed from Oxford, equipped with smoked glass lenses, by which he would measure the position of the stars during the eclipse, along with several pounds of photographic equipment (the war had severely limited the availability of instruments; indeed, Russia still sat on the telescopes it had seized in 1914).

A second British eclipse team traveled to Sobral, Brazil, with two 'scopes, one commissioned from the Royal Irish Academy and featuring a nineteen-foot focal length.

Eddington had to know that none of these fragile instruments, hauled many miles through salty, freezing air aboard ship and then erected in the wild, in extreme heat, could test with any certainty Einstein's precise deflection values. Not only would weather skew the lenses, but even in the best circumstances starlight jittered unpredictably through Earth's soupy atmosphere. Whether Eddington, in his zeal to support his German colleague, would admit these problems to himself (or anyone else) was another matter.

One distraction he *wouldn't* face: fourteen years earlier, Principe's authorities had eradicated the tsetse fly. Plenty of mosquitoes remained; warm rains and springtime humidity meant he'd work under heavy netting, but the absence of biting flies would be a godsend. Now, if there were no revolutions (always a major concern in a colonial outpost), he might be able to redraw the sky.[5]

Clouds curdled the astronomers' view on the day of the eclipse. Eddington obtained two usable photographic plates, though the stars were blurred. The Sobral team also smudged its shots. The telescopes fell out of focus; the observers didn't have enough boxes of ice in the jungle (supplied to them by a Brazilian meatpacker) to keep their chemicals cool for developing the glass plates as carefully as they'd hoped. Initially, in letters to Dyson and his mother, Eddington expressed disappointment.

Yet after several months spent reexamining the results, he declared the cloud cover a blessing in disguise, "since the sun's rays could not seriously affect the [telescope] mirror by heating it."[6] He claimed he now had "complete confidence" that his numbers—though they varied from the Sobral values, which he decided to dismiss—supported Einstein and heralded what the *Times* of London would soon call a "REVOLUTION IN SCIENCE."[7]

"Light caught bending," said the *Daily Mail*. "All Askew in the Heavens."

"New theory of the universe. Newton ideas overthrown," the *Times* announced next to another bold line: "The Glorious Dead, Armistice Observance."

"It is the best possible thing that could have happened between England and Germany," Eddington wrote Einstein. Einstein answered immediately, expressing his profound thanks—wincing as he sealed the envelope. At that moment, right below his apartment window, anarchic concussions sounded in the avenue.[8]

Later, when asked why Einstein enjoyed such a tremendous public reputation, Ernest Rutherford replied, "The war had just ended, and the complacency of the Victorian and Edwardian times had been shattered. The people felt that all their values and all their ideals had lost their bearings. Now, suddenly, they learnt that an astronomical prediction by a German scientist had been confirmed [on] expeditions . . . by British astronomers. An astronomical discovery, transcending worldly strife, struck a responsive chord."[9]

Eddington had carefully prepared his announcement. In interviews with sympathetic journalists, before embarking from Liverpool, he'd played up the romance of eclipse expeditions, building suspense for the expected results. He had described Einstein's theories in colorful terms: "A ray of light should be deflected like a bullet when passing an object exerting gravitational attraction"; "On this occasion everything is to be subordinated to . . . *weighing light*."[10]

His initial concern over the blurry photographs only heightened the occasion's excitement. He joked with colleagues: who knew what might have happened to the universe if the tsetse fly had not been eliminated from Principe?

At the joint meeting of the Royal Society and the Royal Astronomical Society in London on November 6, 1919, Eddington crowed that Einstein's "*law* of gravity" had been confirmed. Dyson backed his results. J. J. Thomson, president of the Royal Society and one of the world's first particle physicists, rushed the meeting to a close, thanking the "Astronomer Royal and Professor Eddington for bringing this enormously important discovery before us," though he conceded that "it is difficult for the audience to weigh fully the meaning of the figures that have been put before us. . . . No one has yet succeeded in stating in clear language what the theory of Einstein's really is."[11]

Of the meeting that night, A. N. Whitehead said, "The whole atmosphere of tense interest was exactly like that of the Greek drama." It was, moreover, a victory for "Jewish science" (though that year, the RAS elected not to award its Gold Medal, its highest annual honor, to anyone at all rather than give it to Einstein, the obvious choice).[12]

It was the return of Dante's hypersphere.

In short order, an "Astronomer's Drinking Song" made the rounds of Society gatherings, commemorating the moment:

> We cheered the Eclipse Observers' start,
> We welcome them returned, Sir;
> Right gallantly they played their part,
> And much from them we've learned, Sir;
> No pains nor toil they thought too great,
> Nor left ein stein unturned, Sir;
> Right heartily we asseverate
> Their bottle a day they've earned, Sir.[13]

Yet Eddington did not go unchallenged. In the last few minutes of the November 6 meeting, a man named Ludwig Silberstein rose to speak. He had authored a book on relativity and considered himself, along with Eddington, one of the few men in the world who understood Einstein's ideas. He cited John Evershed's work at Kodaikanal as reason for caution. So far, there had been no concrete evidence of a solar redshift. The eclipse results were just one instance, and the numbers were murky at best. Silberstein pointed to a painting of Isaac Newton hanging, tilted, on the meeting room wall. "We owe it to that great man to proceed very carefully in modifying or retouching his Law of Gravitation," he said.[14]

Eddington responded with derision. Silberstein pressed on, insisting that "if we had not the prejudice of Einstein's theory we should not say that the figures strongly indicated a radical band of displacement."[15] He repeated Kodaikanal's negative results. Eddington wished he had never heard the name John Evershed.

18

Wallal

"Preposterous," Mary said when John explained his theory (she reported her reaction in a letter to his sister Kate).[1] In his continuing efforts to resolve the redshift question, he had noticed another odd phenomenon: a redshift, all right, in solar prominences at the sun's limb, but not the kind Einstein had predicted, and much larger in value. It appeared to John to be a Doppler shift, an indication that perhaps the Earth had some repelling effect on energy radiating from the sun. Mary thought this crazy. John didn't like it either, but he found no other answer for what he'd witnessed. At Mount Wilson, Charles St. John set to work on the problem, recognizing, he said, the "inadequacy" of Evershed's equipment.[2]

John proposed photographing sunlight reflected off the face of Venus, to see if he could detect, from another source and a different direction, the "Earth Effect." Mary worried about his state of mind. His frustrations with Kodaikanal's dated astronomical tools were growing; St. John and others were surpassing his observational capacities using far better instruments. John remained skeptical of Einstein's theories, despite Eddington's campaign to secure their acceptance, but he wasn't sure he had the means, at Kodai, to add useful data to the worldwide debate.

In 1922 another opportunity would appear for measuring, more accurately, Einstein's twisting light: a total solar eclipse, lasting more than five minutes in some places, spreading from Abyssinia to Christmas Island, across Australia, and snuffing out in the far Pacific. Beta Virginis, third magnitude, would be the test star. More than a year in advance John made inquiries of the Indian government and the Royal Society about the possibility of funding a serious eclipse expedition. This could be his chance to make a major contribution to astronomy. As one journalist wrote at the time, if Eddington's 1919 results were affirmed, "We may have to get used to all sorts of queer ideas, beside[s] crooked beams of light in empty space. . . . We may get to talking about the curvature of time . . . atoms of energy, four dimensions . . . and a finite universe. We may be called upon to . . . conceive of arrows that shrink and bullets that get heavier the faster they travel; of clocks that go slower the faster they travel, and of a future that turns back and tangles itself up in the present."[3]

The Indian government informed John it could offer little support. Frank Dyson helped, loaning him the Royal Society's sixteen-inch coelostat for use with his cameras, yet tests with it "were anything but satisfactory both as regards the mirror and the driving mechanism," John reported. "When stars were photographed [with it] there was shown to be a very marked periodical error. . . . The driving screw was found to be worn and out of truth, and the teeth . . . injured in many places through the wear and tear of previous eclipse expeditions."[4]

John secured meager public funding for travel expenses, but his decision to establish base camp in the Maldive Islands proved untenable. "We underestimated the difficulties of transport from Ceylon to the Islands," he noted. "Native craft trading between Ceylon and Mali [were] said to be impossible . . . and the government of India . . . was unable to provide a vessel for the work." He looked into renting a trawler but was advised to avoid small boats: the monsoons had begun. Finally, William Campbell of Lick Observatory in San Jose invited John to travel with his team to the northwest coast of Australia.

Limited resources forced John to consider taking only one other assistant with him, besides Mary. When the man who'd volunteered to join him, a Professor Maclean of William College, Bombay, became seriously ill, John feared he'd have to scrap his plans, but he pushed ahead anyway. He ordered a new nine-inch mirror to strengthen his equipment, but it did not arrive. He didn't have time to calibrate his coronal spectrograph before setting sail. He quipped bitterly to Mary that, "to say the least, my dear," the trip was not beginning auspiciously.

They left India from Madras on July 28, 1922, bound for Broome, Western Australia. Along the way, in Singapore, they both got sick. Mary recovered quickly—she was as fine a traveler as ever: strong, steady, uncomplaining. For several days, John suffered vomiting and chills, and he lost his voice temporarily. While waiting to sail again, despite his weakness, he bought timber and cement to construct a sturdier mount for his camera. Mary implored him to relax, slow down—to no avail. The one thing that cheered him, leaving Singapore, was the name of their ship, the *Minderoo*. He joked with Mary ("Mindie") that the ship had been built in her honor.[5]

The American eclipse team was scheduled to gather with John and Mary and a klatch of Canadian observers near a long, flat beach named Wallal ("Abundant Water"). Lieutenant-Commander Harold Leopold Quick of the Australian Navy would pilot the astronomers from Perth to their camp aboard a schooner called the *Gwendolyn*. He informed them that the spot was so remote, it did not appear on Admiralty charts, but he could locate it using a post office directory. Local shepherds knew the area: every year, they transported barrels of sheep's wool on schooners setting sail from this barren coast. Briefly, months earlier, the Royal Astronomical Society had considered sending a well-equipped eclipse team here but had rejected the location after learning it was the most isolated telegraph-repeating station in the Commonwealth, plagued by flies and mosquitoes year round. The coast was known for cyclones and sharks.

Fig. 22. Landing at 80-Mile Beach, Wallal, Australia, August 18, 1922. Courtesy Special Collections, University Library, University of California Santa Cruz, Lick Observatory records.

Corpses regularly washed onto the beach following sudden storms. An RAS report concluded that the eclipse would reach Australia at "a hopeless point of the coast and [would strike] into the great desert. There are no facilities for landing. The desert is inaccessible, except to camels. There are no railways within hundreds of miles and cars are out of the question."[6]

Here, in brutal sunlight, amid light loam and sparse wattle trees, Mary and John alighted with sixty tons of astronomical equipment, on the afternoon of August 18. A twenty-six-foot tide churned a heavy surf, tossing the boat from wave to wave. "The white population [at Wallal] . . . appeared to consist of six men," William Campbell wrote later. "Of the aboriginal blacks there were several scores attached more or less informally to the . . . sheep station." He added, "As to the future of the aboriginal

Fig. 23. Unloading telescopes in the surf at Wallal. Courtesy Special Collections, University Library, University of California Santa Cruz, Lick Observatory records.

race, it was interesting to note that, amongst the forty or fifty black men and women, there were only two or three children in evidence."[7]

The Aborigines, particularly a father-son team named Moses and Tommy, waded into the foam to help the visitors unload their equipment and stack it onto rickety donkey carts. The waves carried several crates away, which subsequently crashed on the beach, knocking mirrors wildly off kilter. Navy men rowed Mrs. Campbell and some of the other women ashore in small lifeboats. Mary shucked her shoes, hiked up her dress, and stepped into the ocean. She fell and floated in circles. The sea was as warm as bath water. She choked on salt. As she struggled to stand, she noticed onshore black women settling palm-leaf baskets into clumpy, wet sand, and picking up mussels.

In the next several days, she tried to talk star lore with the Aboriginal women, but their awareness of the world appeared strictly limited to seashells and sheep. Mary gathered many lovely shells with their help. When John wasn't erecting equipment he followed butterflies, but his chases always ended in futility. He couldn't run well in the sand. His heels kicked up little comet trails.

He was frankly envious of the Americans' instruments. They had mounted an "Einstein Camera," a fifteen-footer, on a contraption they called the Tower of Babel. The polar axes of their telescopes had been fitted with ball bearings, eliminating troublesome gears. "Dr. Campbell has a huge 40 foot telescope piercing the blue sky above the trees and [it] looks quite beautiful against the red glow before sunrise," John wrote his sister Kate.[8]

About a week before the eclipse, he ran his first field test with the coelostat. "The result was that all hope of getting perfect plates had to be abandoned," he scribbled in his notes. "With the mirror set to the hour-angle at which totality would occur, marked astigmation appeared in the star-images," a side effect of the faulty drive-mechanism and the worn screws. It could not be fixed. Additionally, his camera was tough to focus, for "the lens was sensitive to temperature change." He mounted two spectrographs on packing cases, but "innumerable flies attack[ed] one's eyes the moment an observation was attempted." Dust scratched the telescope's eyepieces. On top of everything else, the shaded glasses he had brought to wear did not seem strong enough. Mary grew more and more concerned for him; his agitation was severe. A local airplane pilot offered to fly the astronomers over vast stretches of the desert. Mary wanted to see the sights, but John warned her about these daredevil pilots—they had a reputation for dangerous stunts, he said. She understood he didn't want her to leave him alone. They did "motor" eleven

Fig. 24. Mary (second from right) with Elizabeth Campbell (at left) and unidentified women at Wallal. Courtesy Special Collections, University Library, University of California Santa Cruz, Lick Observatory records.

miles inland together, through the "weird Australian bush," to watch kangaroos standing "on their hind legs and tail, their little hands held up in front—a most astonishing spectacle, like a scene in mezozoic [*sic*] times," John reported to his sister. He spotted a "great big bird like a crane," but he "dared" not approach it as it "had a fierce red eye and an enormous beak about eight inches long."[9]

Back at camp, Moses and Tommy kept the tent areas cool by hauling in cartloads of heavy red sand, spreading the sand on the ground, and moistening it with seawater carried in kerosene cans. A Mr. Rhodes of the Australian Navy awakened the astronomers each morning at dawn with blasts from his trumpet, and he provided tea for them every afternoon, though these

Fig. 25. Elizabeth Campbell at a telescope, Wallal. Courtesy Special Collections, University Library, University of California Santa Cruz, Lick Observatory records.

breaks were unpleasant (even treacherous), with winds often reaching thirty miles an hour.

Using a meridian instrument brought from Lick Observatory, Campbell's team discovered that "the moon in its orbit was ahead of the position predicted by the Nautical Almanac. . . . [We] estimated that the eclipse . . . would occur some 27 or 30 seconds earlier than the predicted time. We were accordingly on our guard against a surprise of this sort," he noted in his report.[10] It is doubtful that he shared this information with John (the competition to test Einstein was intense, even among friendly colleagues). John later admitted he was flustered by the timing on eclipse day: "It took the whole of the last minute of totality to get the [spectroscope] slide safely closed and removed from the camera: I thus lost my chance of getting a good view of the corona with binoculars, as I had intended."[11] Also, he had fumbled the first exposure while Mary counted out the seconds.

The day of the eclipse brought a spectacularly clear sky. An observer from the Western Australia Astronomical Society wrote, "As totality approached, the fowls retired to their roosts, and the sheep, horses, and cattle came from under the shelves of the trees and commenced to feed, just as they are accustomed to do in the cool of the evening at sunset. The flies . . . were much affected by the changed conditions, and appeared to become quite paralyzed, so that one could pick them up as if they were dead."[12]

The temperature dropped eight degrees. The horizon turned yellow, then dark blue. Purple shadows brushed the sand.

Carefully, John placed his spectrographic slides in a canvas bag. That night, in a tent kept cool and dark by a netting of wattle leaves, he developed two of the plates. They showed a "considerable amount of fog and other defects," he wrote. He decided to wait until the teams returned to Broome to develop the rest of the slides under better conditions. Mary could feel disappointment rising from him like a layer of heat.

On the scheduled departure day, September 24, a massive wind storm raised the surf. Efforts to reembark were abandoned. The following day, the Aborigines loaded equipment onto a series of lifeboats and began to row toward the schooner anchored just offshore. The wind tossed one of the boats end over end, soaking the Einstein Camera. It took over twenty-four more hours for the astronomers to finally shove off. They left at midnight, when at last the waters were calm. Two days later than scheduled they arrived in Broome.

In town, the chief inspector of police arranged entertainment for them: a beef feast and a *corroboree*—Aboriginal songs, dances, plays. A local ice-making company supplied the scientists with blocks of ice made from rainwater so they could develop more plates. Miserably, John noted in his annual report, "All were found to have failed for one reason or another ... movement of the star-images, poor definition of the corona due to the bad driving [mechanism]." Other pictures were "in some unexplained way ... fogged over."

He concluded: "This completed the failure of our eclipse expedition."

He wore a stoic face. Mary saw the effort he was making in his clenched jaw. In a letter to Kate, he tried to remain matter-of-fact: "I think Mindie told you the eclipse was a total failure. The sky was perfect, but I had very bad luck and got nothing at all. So there is a year's work thrown away. I had great difficulties to contend with from the beginning. . . . I think this will be my last eclipse expedition."[13]

He saved the most excoriating remarks for his year-end report to the government:

> A considerable amount of risk is inevitable in all eclipse expeditions, but it is usually associated with the chances of fair weather. Failure under the ideal conditions of a perfectly clear sky, with excellent definition, and a long duration of totality, is deplorable, especially when public funds have been risked. . . .
>
> If British manufacturers could be induced to abandon the old methods, and apply ball bearings to all moving parts in astronomical instruments, as should have been done thirty years ago, an enormous gain would result in the uniformity of movement so essential in this research.

On April 12, 1923, William Campbell cabled Albert Einstein to say his team's Australian results had confirmed the bent-light prediction. Campbell assured the physicist no further eclipse tests would be necessary. Einstein would be catapulted to international fame: "Men of science more or less agog over results of eclipse observations," reported the *New York Times*.[14] Campbell (and Eddington) would be credited with a major contribution to the field of astronomy, while John would be a footnote.

19

Departure

The Eversheds' equipment arrived safely back in India and made it up the hill on ponies, but John's enthusiasm for the work he did at Kodai never really revived. "Even if we had been successful [in Australia] I was afraid he might be overtired and feel the reaction after [everything] was over," Mary confessed to Kate in a lengthy letter. "He was wonderfully well, and always hungry, in Wallal, in spite of the hard work every day, and the heat and tiresome flies." And then, after the disappointment, he acted "so fine over it all." This worried Mary. How much pain was he hiding? "He just said nothing about his results unless he was directly asked, and then told quite simply the truth about it." Colleagues in England and India assured him he was not to blame for the debacle, Mary said—"his work at Kodaikanal during all these years needs nothing added . . . [and] the Indian government who sent him are responsible for the risk of failure, because the workshop facilities here are out of all proportion feeble for the importance of the observatory, and especially when compared with those of the American[s]."[1]

The couple returned to seven weeks of stacked-up mail, a flooded kitchen (heavy rains had cracked their brick chimney), and several undeveloped sunspot photographs awaiting analysis. John resumed his redshift studies—he was still not prepared to accept all of Einstein's theories. It cheered Mary to see him "com[ing] down to lunch so happy because he had taken some absolutely perfect spectra" one day, and she engaged him in reading aloud with her, at night, H. G. Wells's *Outline of*

History, the story of the Earth from the Stone Age to the first civilizations.[2]

His mood worsened again when an "untimely burst of monsoon" halted daily photography and he began to suspect the accuracy of his spectrographic measurements. His equipment was just too limited.

Finally, Mary wrote Kate, he decided "his Venus observations show there is no difference between the back and front of the sun. . . . I am so thankful that a definite result has at last been reached. Venus has been so elusive, luring him always on and on and giving no definite answer to his question. . . . And I am glad too that we need not believe in an Earth-Effect on the sun, for it always seemed preposterous."

John told her he thought he had achieved all he possibly could at Kodai. He asked if she'd contact an uncle of hers in Gomshall, back in England, about finding housing for them there. He dreamed of building a private observatory in his yard and continuing his solar studies informally without administrative pressures.

By late February 1923, the couple was packing—Mary her books and her old notes on Dante, John his instruments. She sold most of their crockery and furniture, including her piano, to British expats in Tamil Nadu, and found a pair of "devoted dog lovers" to adopt sweet Remus.[3] Among the observatory staff she disported her tennis bats. She wondered how the change in her circumstances with John would alter their remaining years together.

Despite his weariness, John confessed to Kate,

> It is as hard to leave this country as it was for me leaving England: we have grown so fond of the sholas, and the grass, and the hills and cows, and wild animals, and all the different kinds of butterflies. Also there are many friends,

both Indian and English, that we will be very sorry to part with, American too. . . . The Indians love a little tamasha [celebration], so we spent the afternoon having one. The first assistant, fourth assistant, and the meteorologist all made little speeches. . . . The subordinate staff, consisting of peons, lascars, dark-room boy . . . had their meal on the floor. . . .

After I had thanked them in words that I hope were suitably chosen, we all had tea together. It is quite an

Fig. 26. Mary and John (center) with Assistant Director T. Royds (to Mary's right) and his wife, along with the Kodaikanal staff. Courtesy Science Museum Library / Science & Society Picture Library.

> unusual thing for Brahmins to have a meal with Europeans, but they made an exception in this case, and their wives had made little dainties for us, and we had quite a little friendly talk, the wives of course not there. They are very shy birds.[4]

Krishna Aiyer told John the observatory would never again be blessed with such a fine director.

The sky was unusually blue and clear during the couple's last days on the hill, and the garden had never looked more splendid. Friends told Mary an early spring awaited her in England; she wrote Kate that "Jack is already in imagination reading a paper before the R. A. S. on his . . . spectrum measure of Venus, the most reliable that has ever been made. . . . I hope the series of spectra, which will soon be finished now, will come up to his hopes."[5]

In fact, despite his eagerness to reembrace "Jolly Old," John worried that "of course the Anglo-Indian when he comes home finds himself a nobody. The dignity of office has departed, and his brothers and sisters think of him as he was when a child, and he no longer has servants who think of him as a god."[6] It is notable that he called himself a derogatory name here—"Anglo-Indian." He knew that, in class terms, because of his long years away he would remain a partial exile, even back in England.

In early May, on the Eversheds' final trip from the observatory to the harbor where they'd board the liner *Lancashire* for home, a tiger bounded across a back road in front of their car: the first wild cat they had seen in sixteen years in India.

20

Who's Who in the Moon

On Moon Hall Lane, at Pitch Hill in Ewhurst, Surrey, in a quiet area called Highbroom, John set up a private observatory, half underground with a metal dome on top. For the next thirty years he continued his observations of the sun.

He was elected a Fellow of the Royal Society and awarded the gold medal of the Royal Astronomical Society for his investigations of radial motion in sunspots. By now, the RAS had opened its doors to women, and Mary was elected to serve on its library committee. For the British Astronomical Association, she organized a Historical Section "to study the history of astronomy and to co-operate in research, helping to bring new facts to light and unearthing facts now buried in old books and papers."[1] At the time, academia did not recognize the history of science as a formal branch of knowledge.

At a gala dinner in the restaurant Frascati near Piccadilly Circus, the BAA fêted the Eversheds after their "long absence in India, an event that evoked much pleasure among members," said a BAA report.[2] Mary, usually indifferent to high society, had always loved the group's annual dinners at Frascati—the restaurant's gold porticoes, the lavish red-pile carpets and chandeliers, the wall-length curtains. On the menu: Hors d'Œuvres Assortis, Turbuton Florentine, Pommes Nature. The organization had become much more substantial since Mary's paper on the Aborigines, nearly thirty years earlier. Regular speakers at the meetings now included Arthur Eddington, Frank Dyson, and the film comedian Will Hay—an avid amateur astronomer when he wasn't clowning on the screen in the manner of his grumpy old idol W. C. Fields. In 1933, Hay would discover the Great White Spot on Saturn.

The Eversheds' joyous homecoming was shattered by the unexpected death of John's sister Kate, after a brief bout of pneumonia in the spring of 1923. "She was one of the most lovable people [I have known], and her life was spent for others," Mary reminisced in a letter to John's brother, Harry. In an age when "people are often apt to be critical and scornful," Kate took the time to "see the good in everyone ... and so to help everyone to be their best," Mary said; her passing was an awful blow, but "we must think that it is far happier for her to have had so short an illness ... than to linger on into old age and feebleness."[3]

Mary reminded John that Dante had written in *The Divine Comedy* that when a person dies, the soul dissolves its bond with the flesh but retains *in virtute* all its faculties—the sensitive powers become muted, but the rational powers, memory, intelligence, and will, manifest more keenly than before. The soul imprints itself on the air as a result of its innate virtues, and therefore creates a *forma novella*, a rainbow following the spirit that spawned it. Mary was pleased to think of Kate now as a gorgeous spectrum.

John was so upset he could not discuss his sister. "So poor old Kate has gone home," he jotted to his brother, and that was all.[4] Aside from Mary, Kate had always been his staunchest supporter. At least Mary was relieved to see how the English countryside—the fresh strawberries, so much better, to her taste, than any fruits they could find in Kodai—had improved John's body and spirit.

Her own health seemed rather fragile, perhaps because she had been a relatively young woman when she had sailed for India and she returned to England so much older. As literary-minded as ever, she recalled Odysseus's arrival in Ithaca following the Trojan War, dressed in ratty old beggar's rags. She remembered Theoclymenus delivering to Penelope's suitors the prophetic news of their fate: "The Sun has been obliterated from the sky, and an unlikely darkness invades the world." (Mary also recalled from her readings that Plutarch believed this passage in Homer described a total solar eclipse—astronomers had calculated that an eclipse occurred over the Greek Islands on April 16, 1178 B.C.).[5]

It was striking to her how much dimmer sunlight appeared in the northern climes than in southern India. Often, she found herself rubbing her eyes. What was it Virgil said to Dante? "Because you still / have your mind fixed on earthly things, / you harvest darkness from the light itself."[6]

In her Victorian country house, she hung new mirrors in the bedroom to augment the splendid linens she had brought from Tamil Nadu. Her face was drawn and pale as she did her hair in the mornings. Aristotle had said, "If a woman chances during her menstrual period to look into a highly polished mirror, the surface of it will grow cloudy with a blood-colored haze."[7] But Mary knew she needn't fear her mirrors; she had weathered menopause in Kodai. More challenging were her afternoon walks through the forested neighborhood, for her bones had grown brittle. She knew the area around her house had been prosperous since the medieval period. This pleased her. The neighborhood still featured a number of smoke bay houses, stone cottages, and old cedar barns, turned now into private residences.

At night, while John sat in the yard behind the rose garden aiming his 'scope at the sky, his hand-built dome turning with the stars, Mary read in bed. As usual, Dante. Virginia Woolf as well: "So with the lamps all put out, the moon sunk, and thin rain drumming on the roof a downpouring of immense darkness began. . . . But what after all is one night?"[8]

Recently, Mary had discovered the American poet Emily Dickinson through Dickinson's acquaintance David Todd, an astronomer at Amherst University, a veteran of eclipse expeditions, and an admirer of John's work (he had referenced John's research in his 1922 book *Astronomy: The Science of the Heavenly Bodies*). Todd's wife, Mabel, had edited Emily's poems. Certain poems' imagery, of the poet staring heedlessly at the sun and of trying to inhale the whole night sky, spoke keenly to Mary.

For Virgil's two thousandth birthday in 1930, Mary delivered a paper to the BAA on Dante's debt to the *Georgics*. This sparked in BAA members a desire to mark astronomical anniversaries, which led to the founding of the Historical Section, with Mary as director. One of her international colleagues, Professor Pio Emanuelli of the Vatican Observatory, sent Mary a framed portrait of Atlas bearing a star-sphere on his shoulders to mark her new initiative.

The section gained an international reputation for investigative thoroughness based on Mary's research. A Dr. Stokeley of Pittsburgh wrote to her, saying American astronomers had long scratched their heads over the source of Carlyle's statement that he had never been taught to recognize the constellations. Could the Historical Section trace the quote? It turned up in an 1865 magazine, *The Leisure Hour*. Carlyle had written a letter to the editor. "For many years it has been one of my constant regrets that no schoolmaster of mine had a knowledge of natural history," he said. "Why didn't somebody teach me the constellations . . . and make me at home in the starry heavens, which are always overhead, and which I don't half-know to this day?" To which Mary remarked: "Had Carlyle lived just a little longer, surely he would have been delighted to join the BAA!"[9]

The Historical Section embarked on its most ambitious endeavor in 1933. At the association's February meeting that year, questions arose concerning the small moon crater Cichus. A member asked, "Who *was* Cichus? What did he do? When did he live? And did he really deserve a place on the moon?" No one knew the answers. Over the next several months Mary employed her research skills to discover that Cichus was the Latinized name of Cecco d'Ascoli, an ancient astronomer and necromancer. "However, the general problem of lunar nomenclature, and of a great many equally puzzling lunar names was not solved," Mary wrote. She suggested compiling a *Who's Who in the Moon*.[10]

At the start of her task, to place herself in the proper imaginative mindset, she reread Dante's cantos on the moon. There, Beatrice assumes the role of a strict scientist, conducting thought-experiments with mirrors and flames to teach Dante the lunar nature—anticipating, by nearly three hundred years, Galileo's tests of light.

In her introduction to *Who's Who in the Moon*, Mary wrote that many of the ancient names on the "spotty globe" are "not explained; the givers having apparently forgotten that persons well known in their time might not be so in ours. . . . An historical and often romantic interest . . . attaches to the obscure and forgotten names, and to track them down has been a lesson in astronomical history and sometimes quite an exciting chase!"[11]

She learned that the first official guide to list names of craters was published in 1645 by Michael van Langren, or Langrenus, of Brussels (1598–1675). Since then, many names from many cultures had been added to the map. Two examples, annotated by Mary:

> BABBAGE. Charles Babbage, 1792–1871. The inventor of an analytical calculating machine. He . . . introduce[d] in England the notation of the differential calculus.

> ZUPUS. Giovanni Battista Zupi, c. 1590–c. 1650. Italian Jesuit . . . the first to see Jupiter's belts. . . . He was probably the first also to see the phases of Mercury clearly; they had been suspected by Marius, Hortensius, and Galileo.[12]

A 1939 review of *Who's Who in the Moon* in the *Journal of the Royal Astronomical Society of Canada* praised its "most interesting details." "Any astronomer, professional or amateur, will enjoy this publication and would find it a useful book to have continually on a convenient shelf," the reviewer wrote. "Mrs. Evershed is to be congratulated on bringing the labours . . . to such a happy conclusion."[13]

The chore had not been easy. Once again—as with *Dante and the Early Astronomers*—politics had impeded Mary's progress. International war led to paper shortages, causing erratic shipments of documents and proofs.

She would have been pleased to know that, after John's death, a small impact crater on the moon's far side would be named for him in honor of his solar work—like John, it is narrow and long and somewhat worn.

21

The Maunder Minimum

It was not until the 1960s that Einstein's bent light was accurately confirmed, with the advent of radio astronomy and the discovery of the quasi-stellar radio source 3C279, close enough to the ecliptic to be skimmed by the sun and measured by very-long-baseline interferometry. Still, when Charles St. John said in 1923 that he had at last achieved redshift results at Mount Wilson convincing enough to support Einstein, major questions about general relativity seemed to be settled. "The conversion of Mr. St. John is of obvious importance, and . . . leaves the matter now in no reasonable doubt," *Science* magazine wrote.[1]

In December 1923, John complained in a letter to a friend that the "general euphoria" over St. John's announcement, affirming William Campbell's Wallal findings, would erase the fact that *he* was the first astronomer to obtain possible glimmerings of the redshift, but no one had taken him seriously because they considered his instruments inferior. (John failed to mention that, at the time, even he had scorned the Kodai numbers.)[2]

When Einstein visited Mount Wilson in 1931, St. John's reputation as the world's leading solar spectroscopist locked in. John retreated to his backyard observatory.

By the early 1930s he'd become engaged by a problem raised by his old friend Walter Maunder. Convinced, in part, by Mary's insistence on the value of history, Maunder had been studying historical records of solar activity. He had uncovered a clear pattern: from 1659 to 1717 there had been a dramatic "flattening" of sunspot eruptions; this dormancy appeared to dovetail with a period that meteorologists called the "Little Ice Age."[3]

"[Only a] few stray spots [were] noted during 'the seventy years death'—1660, 1671, 1684, 1695, 1707, 1718," Maunder wrote in the *Journal of the British Astronomical Association*. "Just as in a deeply inundated country, the loftiest objects will still raise their heads above the flood, a spire here, a hill, a tower, a tree there . . . so the above-mentioned years seem to be marked out as the crests of a sunken spot-curve."[4]

His discovery had been augmented by the independent work of a man named Andrew Ellicott Douglass. Douglass had been studying plant life in the American Southwest. He wanted to know why large trees grew in Arizona's highest altitudes but only smaller trees peppered the lowlands. Eventually, his investigations led to a publication that caught Maunder's eye: *Climatic Cycles and Tree Growth: A Study of the Annual Rings of Trees in Relation to Climate and Solar Activity.*[5] Douglass reported that sequoias and yellow pines reflected eleven-year sunspot cycles in their inner rings. According to the trees, a major absence of solar activity occurred between the 1650s and the 1720s. He believed the sun had a major impact on precipitation, violent storms, and weather systems. Maunder concurred—and went a step further, suggesting Earth's magnetic poles, its crop production, and maybe even human history responded strongly to solar cycles as well.[6]

As evidence:

Only thirty-five years after Galileo had cataloged the "impurity" of sunspots, people throughout Europe were gazing through portable telescopes, asking, "Where *are* those damned spots?" In November 1623, Galileo planned a winter trip to Rome but was forced to cancel because of exceptionally cold weather. Shortly afterward, Switzerland experienced an unusually chilly, wet summer. Grapes were harvested several months late and heavy snows covered the Alps in August. Not long after that, unseasonably cold rains destroyed the olive crops in Crete.[7]

Civil war flared in England, the Thirty Years' War erupted on the continent, strife spread in China and throughout northern Europe. Famines, along with outbreaks of illness and plagues, compounded these political/religious struggles.

In 1648, British astronomers reported a total absence of sunspots. In early 1649, the Thames River froze solid. England, still suffering a civil war, reported a doubling of the mortality rate in Berkshire between 1642 and 1646.

From February 1653 until January 1660 the French astronomer Jean Picard recorded no spots.

In 1667, the poet Andrew Marvell wrote, ". . . Man to the sun apply'd / And Spots unknown to the bright star descry'd."[8]

Venice's canals iced over in the early 1680s.

Dante's killer, malaria (*mala aria*—"bad air"), spiked during the Little Ice Age. *Ague* is the malady's English name. Shakespeare, whose last days coincided with spotlessness, used the word in eight of his plays. Caliban, cursing Prospero in *The Tempest*, hisses, "[May all] the infections that the sun sucks up / From bugs, fens, flats, on Prosper fall . . ."[9]

And then, after 1700, observers noted a sharp uptick in sun spot activity. Stephen Gray, a British astronomer, saw a "white flash" on the sun in 1703. He wrote, "The [sun]spots being thrown up soe suddenly argueth that they are ejected with a most violent force . . . they seem to proceed from the Internal parts of the sun . . . as sand stones, ashes, etc., are thrown from Mount Etna."[10]

Sightings of the northern lights grew more frequent after 1720. Temperatures rose across the length and breadth of Europe.

In 1726, Jonathan Swift wrote that the "learned men" of Lilliput feared the "face of the sun [would] by degrees be encrusted with its own effluvia, and give no more light to the world."[11]

Between 1710 and 1713, William Herschel had recorded a stretch of "spotless sun," but welcomed after that a "relative abundance" of spots, coinciding in general with a dramatic warming trend.[12]

"Can [sunspots] affect us here on earth," Maunder asked, "even if we hid ourselves in her depths, shut off by many feet of cold ground from the sight of the sun, from the knowledge of day and night, or summer heat and winter cold? . . . Yes. The origin of our magnetic storms *does* lie in the sun."[13]

In Surrey, John turned his new design, a hollow prism filled with ethyl cinnamate, toward dawn's light. He happened to discover that the sun's angular rotation increases with its height above the photosphere, a fact unrelated to the climate debate. He was grateful to Maunder. The sunspot problem took his mind off his Einstein failures, though results were likely to remain inconclusive for decades. In any case, it felt right to cap his career this way. Concerns about the sun and monsoons had led to the establishment of Kodaikanal, and it was Kodaikanal to which his name would always be attached.

Already, his years in India seemed a lifetime ago.

22

The Remade Universe

In 1925, "a new epoch opened" in human history, according to the British Astronomical Association. Since the "time of Ptolemy," astronomical timekeeping—designating noon as 00:00:00—had been the West's default mode.[1] William Herschel had argued that skywatchers preferred noon as the day's starting point: this meant observations made on any single night would fall on the same date. The Nautical Almanac Offices in both Britain and the United States followed astronomical usage, but after the First World War, British seamen lobbied the national hydrographer to switch to civil time, making midnight the day's beginning. This kept the sunlight hours whole during every twenty-four-hour period, rather than splitting them in half. This way, life was easier for boatmen, businessmen, and traders. The Nautical Offices agreed, as did the Royal Astronomical Society, and in 1925 the change was made.

Most of the world's citizens ignored this new cycle. It didn't alter their daily experiences. As always, humanity seemed bent on nasty politics, no matter how events got measured. For example, early in 1940, a BAA member reported to the organization that he had "observed an unusual 'solar eclipse.' While recording details of the solar disc with the spectrohelioscope he suddenly became aware that the sun was being occluded by a sharp opaque curve; for a second he was nonplussed. Here was an unpredicted solar eclipse, and the moon eleven days old! It transpired that the rounded nose or fin of a barrage balloon was slowly encroaching between the sun and the coelostat."[2]

Soon, German Zephyrs were pounding the city of London. Exhaust trails, shrapnel, blazing aircraft . . . the Battle of Britain had begun. Because of nightly blackouts, and for safety reasons, the BAA kept altering its meeting times. The restaurant Frascati was severely damaged in a bombing raid one night, ending the astronomers' festive dinners.

Despite a tendency to tire easily, Mary remained extraordinarily active in the organization throughout this unsettled period, running the Historical Section and arranging talks and meetings. One early-spring afternoon she invited the Reverend T. E. R. Phillips, rector of Headley Parish, a longtime BAA fellow, and friend to several cutting-edge physicists, to address a small group of amateur astronomers in Westminster's Central Hall. His paper was titled "Some Recent Views of the Physical Universe and Their Reaction on Present-Day Thought." It gracefully summarized post-Einsteinian existence. The "universe has grown so much larger than it used to be," Mary said by way of introduction, "and the stuff it is made of—just the same 'ordinary matter' which makes our Earth—has become far more wonderful and more mysterious."[3]

The reverend began by dismissing Dante's era (Mary held her tongue): "For many centuries following the golden age of Greek culture there elapsed a period of almost complete stagnation and paralysis in regard to those matters with which we are now concerned. This arose partly through . . . [the] weight of authority which some of the great teachers of the past—Aristotle especially—exercised over early and medieval thought. . . . [There could be] no emancipation of men's minds from the thralldom of superstition under which the peoples of medieval Europe lived and suffered."[4]

In the seventeenth century, he went on, Galileo had made central the "study of [stellar] positions," and now, following Einstein's contributions, "astronomy has taken into partnership with itself the science of physics. What an amazing difference between the universe as we now conceive it to be and the universe the early astronomers centered on our little world!"[5]

Relativity, he argued, "has completely undermined our former belief in the absoluteness of the familiar standards of measurement. . . . Rods and clocks and scales are not absolute at all, but vary with the motion of the observer relative to the velocity of light." Then he waxed philosophical: "Perhaps the most striking result of modern discoveries is [that they] have combined to increase our appreciation of the significance of mind in our experiences of the external world." In conclusion: "If we hold that God, though transcendent, is *immanent* in the universe . . . He reveals himself in some fashion wherever there are minds with the capacity for knowing Him."[6]

Mary thanked Mr. Phillips for his insights. Then she opened a general discussion. Perhaps it was no surprise that *faith* obsessed everyone—in a group rattled by war and the loss of absolutes.

One audience member rose to rail against the "anti-God Movement" in Britain, blaming "Messrs. G. B. Shaw [and] Aldous Huxley." Another said, "There is a reason for the large size of the solar system. When God decided to give Free Will to men He had to take precautions that that great experiment should not end in disaster. If the solar system had been much smaller than it is, men, in their perversity, might have found means to interfere with its smooth running." Mary's friend Annie Maunder derided what she saw as the medieval mindset—which, sadly, afflicted many of her contemporaries: " 'We neither hear nor see anything with accuracy' . . . This is, indeed, the unforgivable sin in science, the denial of the necessity of observation and of making the facts fit the hypothesis, *not* the hypothesis fit the facts." The day's final comment, made by a speaker in the back of the room, left a chill in the hall: "The Lord God may have plans for the future beyond our comprehension . . . these plans may even extend beyond the duration of the earth as it now is; for we are told [Psalm 102: 25–27] that the time may come when God will lay aside the heavens as a worn-out garment and change His vesture. Modern research seems now to point towards the same outcome."[7]

23

Return to Origins

In light of the uncertain present and a cloudy future, Mary felt a need to return to astronomical origins—as if seeking some unity between Dante's universe and our own. The last formal paper she prepared, for the *Observatory*, was an overview of "Arab astronomy." She praised the Arabs for "restor[ing] to Europe"—and to Dante—"the forgotten work of the Greeks, in some ways disfigured, but in other ways improved by better observations and instruments"; for "tak[ing] up the torch that Greece dropped and hand[ing] it on. They prepared the way for Copernicus to remove the Earth from her central position and place the Sun there[,] . . . for Galileo to discover new wonders in his telescope."[1]

It pleased her to imagine "one of the most romantic chapters in . . . the history of astronomy," when the Arabs adopted ancient constellations and star names, such as Ptolemy's Betelgeuse ("Shoulder of the Giant") and Denebola ("Lion's Tail"), and passed them on to us; it thrilled her to trace echoes of old Arabic words in our current designations—El-nesr el-tair, the Flying Eagle, in Altair from the constellation Aquila, for example—and to find Bedouin names still gracing our sky (Alphard, the Lonely One, in Hydra).[2]

Most of all, it reassured her to think of Jews, Christians, and Moors cooperating in Spain's twelfth-century translation workshops, restoring the *Almagest* and other star books, spreading them across the globe. She imagined Toledo as a magical realm, a city above a whispering river on a hill, a catacomb, a human hive vibrating with pulleys and ropes and gears, mirrors and lenses and dials, shadows absorbing the essence of time, barrels of bath water hauled from the stream, warming slowly on chalky stone roofs, sprinkled with sky-dust. Curled pages of long-lost writing scattered like empty insect shells across the city's tables, floors, and cobbles—translucent, faded, ready to be filled once more.

"It is to be hoped that our star-names will never be banished from our maps and globes, for they link us" with the past, Mary said.[3]

24

Northern Lights

From June 1923 until the pages trailed off at midcentury, Mary's "Visitors" book, lying open on a table in the house on Moon Hall Lane, recorded the names of family members and professionals who arrived to pay their respects to John and Mary: the Maunders and the Astronomer Royal; young researchers from observatories in Cairo, Paris, Greenwich, and on remote U.S. hilltops. Mary Proctor dropped in for literary tips as she was composing her magnificent *Children's Book of the Heavens* (previously, she had profiled John for the British press, noting that his "only fault is the modesty which has prevented his pushing his work into greater prominence").[1]

Mary's nephew A. D. Thackeray, charmed by his aunt's romantic life, had become an astronomer and regularly came to Ewhurst for guidance and advice. In the years ahead, he would serve as chief assistant at the Radcliffe Observatory in Pretoria, specializing in the study of globular clusters and the Magellanic Clouds.

Old friends from Kodaikanal sailed to England just to see the Eversheds, sharing exciting stories of India's fight to lose the British yoke. All across Kodai, the Union Jack had been ripped down and replaced with the national tricolor flag. An old Bharathiyar poem, "Aduvome pallu paduvome," had become an exhilarating regional anthem.

Occasionally, missing the spirit of travel, Mary and John locked up the house for a few months. They booked excursions to Australia, Greece, and isolated spits of England to watch solar eclipses, reliving their very first trips together. Each time, the weather spoiled their views of the sun, leaving them feeling wistful, knowing their chances of sharing eclipse magic were running out.

In September 1942, as London was still clearing the rubble of war, the British Astronomical Association moved into new accommodations in Burlington House, Piccadilly—a space cast off by the Royal Astronomical Society in search of bigger meeting halls. To mark the occasion, the organization asked Mary to write a history of the BAA. She approached the assignment eagerly; shuffling old papers, she was amazed at how swiftly astronomical knowledge had evolved in her lifetime. In the group's early days, people believed that comets caused meteor showers; that canals, deliberately and smartly designed, crossed the rolling red plains of Mars; that flashes on lunar mountain peaks (when pale rocks caught the sun's rising glare) were messages from intelligent life. Space was thought to be empty, endless, and flat.

While Mary revived the past, John, puttering beneath his backyard dome, produced daily spectroheliograms in hydrogen light. He designed ever-larger prisms with greater resolution. In the afternoons, he'd walk into town to the Halls and Company Builders' Merchants to purchase material for making new instruments. There, he met a young woman named Margaret Randall, who worked in the shop's front office as an accountant and receptionist. John was impressed with her knowledge of construction techniques and mechanics. They became friendly. He brought her home to meet Mary. The women hit it off, sharing a passion for gardening. Margaret was fascinated by John's observatory; under the Eversheds' patient guidance, she became familiar with the constellations.

When Greenwich astronomers predicted a spectacular season for the aurora borealis in Stranraer, in the far southwest of

Scotland, the Eversheds invited Margaret to join them and the Maunders on a northern lights pilgrimage. An undated photograph—taken sometime in the late 1940s—shows John, Mary, and several unidentified men standing with Margaret and Annie Maunder on the SS *Strathaird* on their passage north. Annie—Mouse—looks frumpy as she stands, stooped, smiling impatiently at the camera. Mary looks gaunt and tired but keenly alert, standing erect with her back against a cabin doorway, wearing an ankle-length overcoat and big, bulky shoes. Her hair is short, severe, emphasizing the thinness of her face and neck, still long and slender after all these years. John, dressed

Fig. 27. John and Mary aboard the SS *Strathaird*, along with Margaret Randall (in front of John), Annie Maunder (in front of Mary), and unidentified others. Courtesy Science Museum Library / Science & Society Picture Library.

handsomely in suit and tie, stands next to her, gray-haired and distinguished. His left hand, fingers loosely spread, drapes Margaret's shoulder. She is standing in front of him, in a plain light-colored dress, smiling, leaning gently against his chest. She is the youngest person in the photograph.

John's easiness with her suggests that a strong intimacy had developed between them. One can't help but wonder what Mary thought of her young friend's posturing with her husband. Terry Evershed, a distant cousin of John's with whom I corresponded while writing this book, remarked wryly on "that hand!" but knew no family stories concerning when and how John's affections for Margaret grew. We have no indications in Mary's letters about her feelings on the matter.

The *Strathaird* was a white-hulled luxury liner just returning to civilian service after evacuating six thousand British and Allied troops from the coast of France at the height of the Second World War. In Stranraer, the astronomers, hungry and sea-wobbly, hiked past the Castle of St. John, a medieval tower house, and staked out a sky-watching spot on the banks of Loch Ryan.

The lights did not disappoint: just after dusk, a green band crossed the ecliptic. It began to waver like moonlit water in an ice pond in the wake of a minor earth tremor; the waver became a wind-blown drape, heavy velvet, a silence crackling in the night, pulling swiftly across the vast black window of the sky, now magenta, blue, and purple in addition to green, a mosaic in motion. Waterfall mist. Mary gripped John's arm and let herself go dizzy in the dance.

Back home, some months later, she complained of exhaustion—more than usual. Her BAA research languished, and she passed the work on to others after organizing the remainder of the files. At John's insistence she submitted to a series of medical tests. Doctors confirmed it: cancer had swamped her body.

Whatever the long-term prognosis, whatever the time frame, she wanted to stay at home. It calmed her to sit in the yard and meditate on the blue lupines in her garden, to gaze at the afternoon light glinting off John's silver dome. She liked to listen to the radio. Sometime earlier, the BBC had broadcast a lecture on Dante by Professor Cesare Foligno of Magdalen College, Oxford. Mary sent him a "charming letter"—"I am overwhelmed by your kindness and appreciation," he said. "Dante's poetry is so great that I often feel that by speaking about it one gets credit under false pretenses, for such success as one may achieve is due to him rather than to any merit one's efforts may have."[2] The exchange brought her back to the pages of *Paradiso*; her poet's company was as sustaining as ever.

In the first flush of her illness, Mary received a visit from Barbara Reynolds, a teacher and scholar at Cambridge who had, by chance, run across *Dante and the Early Astronomers* in the Cambridge library. "From then on, I was able to explain the astronomical references to my students instead of saying, as T. S. Eliot did, that they don't matter and you can skip them," Reynolds said.[3] Mary's nephew was teaching astronomy at Cambridge Observatory. He met Reynolds and arranged an afternoon with Mary. Although Mary was exceptionally weak, the women spent a pleasant day sharing their love of Florence and Dante's lyricism. They had discovered the "vernacular" in one another, as Dante had said people would do—people who shared a spiritual kinship—a "fragrance" more "perceptible in one than in another, just as the simplest of substances, which is God, is more perceptible in man than in a brute."[4]

As a gift, Mary gave Reynolds her old blue leather-bound copy of the *Latin Works of Dante*, with an illustration from the Lateran Mosaic.

At the time, Reynolds's friend Dorothy L. Sayers was translating *The Divine Comedy* into English. Dante, Sayers said, was a "sealed book" to her countrymen, who were raised on "science and psychiatry and television" and therefore "incredibly lacking in literary background." Reynolds introduced her to Mary's "remarkable" book, which Sayers found to be "quite the best guide available to Ptolemaic astronomy and to Dante's handling of celestial phenomena."[5] When Sayers died in 1957, Reynolds completed the translation. She also arranged to reprint Mary's book with Allan Wingate, a London publisher. The second edition met the opposite fate of the first: students of science ignored it, but Dantists welcomed its unique approach to the poem. Writing in the *Modern Language Review*, Colin Hardie of Oxford said the book had been "unjustly neglected." He marveled at how "well equipped" Mary was to bring poetry and science to the "general public" as well as to "specialists."[6] In her introduction to the new edition, Reynolds noted that Mary's work "belongs to the period of the amateur-scholar," a "phase of culture" long past.[7]

On January 30, 1948, Mary heard on the radio that Mahatma Gandhi had been assassinated in Delhi by a Hindu extremist. The India she had known, the world she had loved, was rapidly vanishing. Albert Einstein, the man who had most extensively changed Mary's universe, said, "It was [Gandhi's] unshakable belief that the use of force is an evil in itself, that therefore it must be avoided by those who are striving for supreme justice."[8]

Paradise. It is madly impure. Dante announces this in the very first lines of his third *cantica* when he praises the "glory of Him who moves all things."[9] At the time, the word "glory" was associated with Christian rituals, but the notion of a Mover, one who "moves all things," came from Aristotle by way of the Moors—thus, to many of Dante's Christian contemporaries, it was anathema.

By mixing terms, images, ideas, registers, metaphors, and conceits, Dante made it clear that extremists of all stripes, who claim perfect knowledge of God in His Heaven, know nothing: we on Earth, in our few, flickering days, see only distortions, as in a hall of curved mirrors. Paradise "cannot be described."

Mary took comfort from these passages in her final days, recalling afternoons at Kodaikanal peering at the pure black sky through the impure instruments of her passion—mirrors, lenses, prisms. It was all beyond telling.

On August 25, 1949, an astronomer from the Royal Greenwich Observatory wrote to John saying he would cancel his scheduled visit in light of John's "great trouble"—his wife's "desperate illness."[10] The man had hoped to discuss with John the latest observations John had published in the *Monthly Notices of the Royal Astronomical Society*, a puzzling "instability" of solar wavelengths. Rather than propose an answer for what he was witnessing, as he might have done as a younger man, John admitted the "difficulty of finding a suitable explanation."[11]

Two months later, on October 25 at the age of eighty-three, Mary died in her bedroom at Moon Hall Lane.

Margaret Randall tended Mary at the end. Whatever arrangements may or may not have existed between John, Mary, and Margaret, Mary was grateful for Margaret's kindness. She gave the young woman her blessing, and in 1950 John married Margaret. She was forty years his junior; she would survive him by nearly half a century.

Until his death in November 1956, John continued to monitor solar wavelengths, keeping an accurate record of dynamics he could never explain.

His doctor considered him an "impossible patient" during the final years of his life. "However ill, [he] would always refuse to leave his underground observatory. I well remember having to examine his chest as best I could while he was engaged in taking photos of the sun. At the age of 75 I found that he had an auricular fibrillation and warned him that he must go reasonably quietly for a period of time. On my next visit, as if to defy his pathology, I found him felling trees with an axe."[12]

In a magazine called *Vistas in Astronomy* in 1955, John published an article entitled "Recollections of Seventy Years of Scientific Work." Without Mary's suggestions, her poetic eye, her corrections of his prose, his sentences lay flat on the page. Of his wife, he said only that she was "much occupied" in India "writing her important work" on Dante. The couple's time at Kodaikanal was "characterized by some interesting and, indeed, exciting events," he says: "A great spot in September, 1909," and a "great sun-grazing comet" on the morning of January 17, 1910.[13]

His silence about Mary in this final summation of his life was yet another example of that which "cannot be described."

At the end of *Paradiso*, when Beatrice takes leave of Dante, we feel more strongly than at any other point in the poem his longing for her—yet their parting remains simple, almost unspoken. Dante offers his heartfelt address: "I know the grace and virtue I've been shown / come from your goodness and your power."[14] Beatrice merely smiles at him, as if too overcome with emotion to express anything. The critic Peter Hawkins notes that, in a rare instance, Dante uses the Italian *tu* here, the familiar, intimate form: "In the poem's 11th hour we have the human face to face moment we have been waiting for. A man talks to a woman he loves."[15]

EPILOGUE

Kodai Dusk

Visitors to the Palani Hills will be treated to a variety of sunsets, depending on the season. Dry summers, increasingly frequent in this era of global warming, create intensely blue skies and make mountaintops crisply visible in the distance. Their peaks are nearly free of clouds. Dusks tend to be sharp, edging toward vivid orange and away from gentle shades of yellow and pink. The light reveals isolated patches of blue-white snow in the crevices of the Pir Panjal Range south and west of valley thickets. Dense autumn snowfalls, mounded rocky domes, appear late each fall in the mountains to the east, among eucalyptus and pine, sometimes deep into the month of October. On chilly, edge-of-winter nights, the dominant quality of dusk-light is difficult to determine: does it originate in the setting sun or the rising full moon?

One can almost hear, in the stillness, the Music of the Spheres—it *does* exist, we have learned, in the heart of a singing black hole 250 million light-years distant: a pressure wave humming B-flat, fifty-two octaves below Middle C.

In September, when for nearly three weeks the general direction of monsoon winds switches from southwest to northeast, clouds disperse and dusk assumes the briskness of dawn: green flashes seem to erupt in the western hills just as the sun's disc sinks—a terrestrial or solar effect? The naked eye cannot tell.

Nor can it adjust itself easily as the first stars emerge in darkness. For night-trackers used to particular constellations, Kodai puzzles, straddling as it does a north-south border. The familiar patterns appear to be tilting precariously on the horizon or occupying the wrong location. It is a sky at odds with itself.

The disorientation of early eclipse observers is echoed in this unbalance. The observers believed prominences might be "pink clouds floating in the moon's atmosphere," Mary once wrote. Or maybe they were "mountains on the sun."[1]

A similar confusion appears to fill Saint Bernard's prayer to the Virgin in the final canto of *The Divine Comedy*. His words underscore the impurities of Paradise. In the saint's exordium, Mary is both "virgin" and "mother," "mother" and "daughter" of her son, "humble" and "exalted." More crucially, infinity exists in the finite. In his final lines, Dante the Poet, recounting his life's journey, switches his verbs to the present: as in the angels' consciousness, future and past converge in the now.[2]

A little over a decade after John Evershed departed India and other British-trained astronomers left the scene, a succession of Indians assumed the directorship of Kodaikanal: A. L. Narayan, Amil Kumar Das, M. K. Vainu Bappu. Today, R. Ramesh oversees the work there. In 1977, Kodaikanal became part of the Indian Institute of Astrophysics.

Descendants of the family that served at Kodai in John and Mary's time still analyze the sun at the observatory, using, in some instances, instruments John designed. Each morning at the solar station, Devendran P climbs a spiral staircase and aims a spectroscope through the shutters of the dome, just as his grandfather did; later, in the 115-year-old processing lab, using photographic chemicals stored in an earthen pot, he will develop his images and add them to Kodai's massive record. "The sun, like stars, has a lifetime of 10 billion years," he explained one day to a reporter from Mumbai. "If you want to know about any small changes, you need to have a large amount of data." As the sun set, he added, "I feel more religious than other people, as I can see that there is a universal power which is controlling everything. . . . I worship the sun. . . . We are the children of the sun. . . . This place is like a temple to me."[3]

His son Rajesh, now in his twenties, speaks eagerly of his desire to extend the family's work, work started here over a century ago by a curious British couple. Much of their world has vanished—not a bad development on balance, given colonialism's corrosive aftereffects—though in the observatory's well-kept library, the card catalog that Mary began is still in effect, and the books' tightly stitched leather covers (work lovingly done by young bookbinders under Mary's guidance) shine warmly among the glass plates stored in wooden cupboards, marking the sun's progress through the years. "The sun is in my blood," Rajesh says.[4] The sun is in the soil of these hills. Earth and sky. In opposition lies unity, understanding: so sang the wandering exile, Dante Alighieri, centuries ago.

Generally, now, Dante's astronomy has sunk once more into obscurity, receiving cursory treatment in academic publications—though the poet continues to be appreciated for the accuracy of his science. In 1979, Mark A. Peterson wrote in the *American Journal of Physics* that Dante's conception of the universe in Canto XXVIII of *Paradiso* is "unbelievably apt" in positing a "three dimensional sphere" with "finite volume" but "no boundary." Dante's genius was to make "verbal arguments which closely parallel[ed] . . . mathematics."[5]

In 2000, Alison Cornish, a teacher of Italian, published *Reading Dante's Stars*, the most extensive study of Dante's celestial imagery since Mary's book. For Cornish, the outstanding feature of Dante's universe is its sensuality, the fact that, according to Dante, God is the "love" that "governs heaven."[6] At the heart of existence is the eroticism of the stars, the dynamics of desire. Mary Acworth Evershed was the first to tie this aspect of Dante's writing to scientific truths. On a hilltop with her husband and several hundred pounds of astronomical equipment, she celebrated poetry and the romance of the stars.

Notes

Preface: The Dawn-Light of Ravenna

1. Laura Fermi and Gilberto Bernardini, *Galileo and the Scientific Revolution* (Mineola, NY: Dover, 2003), 94.
2. Sandow Birk and Marcus Sanders, *Dante's Inferno* (San Francisco: Chronicle Books, 2004).
3. Edward Hirsch, "Summoning Shades," in *The Poet's Dante: Twentieth-Century Responses*, edited by Peter S. Hawkins and Rachel Jacoff (New York: Farrar, Straus & Giroux, 2001), 397.
4. Harriet Rubin, *Dante in Love: The World's Greatest Poem and How It Made History* (New York: Simon & Schuster, 2004).
5. See, for example, Mark A. Peterson, *Galileo's Muse: Renaissance Mathematics and the Arts* (Cambridge, MA: Harvard University Press, 2011); Edward Grant, *Physical Science in the Middle Ages* (Cambridge: Cambridge University Press, 1977); Grant, *Much Ado about Nothing: Theories of Space and Vacuum from the Middle Ages to the Scientific Revolution* (Cambridge: Cambridge University Press, 1981); Grant, *Planets, Stars, and Orbs: The Medieval Cosmos, 1200–1687* (Cambridge: Cambridge University Press, 1994); and William Egginton, "Dante, Hyperspheres, and the Curvature of the Medieval Cosmos," *Journal of the History of Ideas* 60, no. 2 (April 1999).
6. Mark A. Peterson, "Dante and the 3-Sphere," *American Journal of Physics* 47, no. 12 (December 1979): 1031.
7. See, for example, Eileen Pollack, *The Only Woman in the Room: Why Science Is Still a Boys' Club* (Boston: Beacon Press, 2015).

1. On the Hilltop

1. T. S. Eliot, "Dante," in *The Sacred Wood* (London: Methuen, 1920), 168.
2. For an English version of the "Quaestio de Aqua et Terra," see A. G. Ferrers Howell and Philip H. Wicksteed, trans., *A Translation of the Latin Works of Dante Alighieri* (London: J. M. Dent, 1904).

3. Dante cited in M. A. Evershed and J. Evershed, "Dante and Medieval Astronomy," *Observatory* 34 (1911): 440–444.

2.
To the Lighthouse

1. See Mookkiah Soundarapandian, *Development of Special Economic Zones in India: Policies and Issues* (Delhi: Concept, 2012), 168. Residents of Kodaikanal often refer to the region as Kodai. In what follows, I use the names interchangeably.
2. Nora Mitchell, *The Indian Hill Station: Kodaikanal* (Chicago: University of Chicago, Department of Geography, 1972), 97.
3. Dante, *Purgatorio*, translated by Jean Hollander and Robert Hollander (New York: Anchor, 2003), 5.
4. Cited as integral to British girls' education, both at home and abroad, in Elizabeth Buettner, "Problematic Spaces, Problematic Races: Defining 'Europeans' in Late Colonial India," *Women's History Review* 9, no. 2 (2000): 288.
5. Virginia Woolf, *To the Lighthouse* (Orlando, FL: Harcourt, [1927] 2005).

3.
The City of Stars

1. John Ruskin, "Mornings in Florence," in *The Works of John Ruskin*, vol. 23, edited by E. T. Cook and Alexander Wedderburn (London: George Allen, 1906), 848.
2. For details on Dante and the mosaics of Florence, see Ernest Hatch Wilkens, "Dante and the Mosaics of the Bel San Giovanni," *Speculum* 2, no. 1 (January 1927): 1–10.
3. Dante, *La Vita Nuova,* translated by Barbara Reynolds (London: Penguin Books, 2004), 54.
4. Dante, *Paradiso*, translated by Jean Hollander and Robert Hollander (New York: Doubleday, 2007), 707.
5. The clearest expression of Dante's themes here, as Mary understood them, can be found in Alison Cornish, *Reading Dante's Stars* (New Haven, CT: Yale University Press, 2000). See Cornish's discussion of St. Augustine, *De Genesi* 4.30.47, *PL* 34.317, at 131.
6. Cornish, *Reading Dante's Stars*, 130.
7. Thomas Aquinas, *Summa Theologica* 1.63.5, translated and referenced in Cornish, *Reading Dante's Stars*, 132.

4. Poetry and Sunspots

1. Thomas G. Bergin, *Dante* (New York: Orion Press, 1965), 45.
2. Dante, *Paradiso*, translated by Mark Musa (New York: Penguin Books, 1986), 205. [Note: Throughout the text of *Dante and the Early Astronomer* I have made use of several different English translations of Dante, in the interests of the greatest possible clarity and fineness of the poetry.]
3. Cited in Thomas Caldecott Chubb, *Dante and His World* (New York: Little, Brown, 1966), 778.
4. Boccaccio, *Life of Dante*, translated by J. G. Nichols (London: Hesperus Press, 2002), 29.
5. Chubb, *Dante and His World*, 762.
6. Dante, *Purgatorio*, translated by Mark Musa (New York: Penguin Books, 1985), 82.
7. Dante, *Paradiso*, trans. Musa, 271.
8. Ibid., 238.
9. Ibid., 119.
10. Dante, *Inferno*, translated by Mark Musa (New York: Penguin Books, 1984), 260–261.
11. Many variations exist of the story of Hans Lippershey's invention of the spyglass. For one such account, see Fermi and Bernardini, *Galileo and the Scientific Revolution*, 45.
12. Matteo Valleriani, "An Organ Pipe as a Telescope," *Max-Planck-Gesellschaft*, www.mpg.de/7913340/Galileo¬_Galilei_telescope.
13. For an English translation of Galileo's lectures, see Galileo Galilei, *Two Lectures to the Florentine Academy on the Shape, Location, and Size of Dante's "Inferno"* (1588), translated by Mark A. Peterson, www.mtholyoke.edu/courses/mpeterso/galileo/inferno.html.
14. Dante, *Paradiso,* trans. Musa, 19.
15. Galileo cited in Eileen Reeves, "From Dante's Moonspots to Galileo's Sunspots," *MLN* 124, no. 5 (2009): 199.

5. "Black Star-Lore"

1. This formulation appears in the *Memoirs of Sir Isaac Newton's Life* by William Stukely, cited in Steve Connor, "The Core of Truth behind Sir Isaac Newton's Apple," *Independent*, January 18, 2010, independent.co.uk/news/science/the-core-of-truth-behind-sir-isaac-newtons-apple-1870915.html.

2. From Mark Twain's *Following the Equator* (1897), quoted in Barry Godfrey, "The Australian Colonies, 1787–1901," *The Digital Panopticon Project*, www.digitalpanopticon.org/The_Australian_Colonies,_1787–1901.
3. John Tebbutt journal entries quoted in Wayne Orchiston, *John Tebbutt: Rebuilding and Strengthening the Foundations of Australian Astronomy* (Cham, Switz.: Springer International, 2016), 66.
4. M. A. Orr, "Black Star-Lore," *Journal of the British Astronomical Association* 9, no. 680 (1898): 68.
5. Notes on an 1898 meeting of the British Astronomical Society, *Observatory* 22 (January 1899): 47.
6. Orr, "Black Star-Lore," 68.
7. Ibid., 69.
8. Ibid.
9. Quoted in Mary T. Brück, "Mary Ackworth Evershed nee Orr (1867–1949), Solar Physicist and Dante Scholar," *Journal of Astronomical History and Heritage* 1, no. 1 (1998): 45.
10. John Tebbutt, Foreword to M. A. Orr, *An Easy Guide to the Southern Stars* (London and Edinburgh: Gall & Inglis, 1897), i.
11. Orr, *Easy Guide*, unpaginated.
12. Dante, *Purgatorio*, trans. Hollander and Hollander, 751.
13. Cited in Barbara Reynolds, "Introduction to the Second Edition" of M. A. Orr, *Dante and the Early Astronomers* (Port Washington, NY: Kennikat Press, [1913] 1969), 18.
14. Ibid.
15. Ibid.

6.
Physical Astronomy

1. Pickering quoted in Dava Sobel, *The Glass Universe: How the Ladies of the Harvard Observatory Took the Measure of the Stars* (New York: Viking, 2016), 13.
2. Ibid.
3. Ibid., 38.
4. Speculation has touched on Mary Somerville, Caroline Herschel, and a schoolteacher named Richard Bloxam, who may have masked himself as a "lady."
5. George Eliot, *George Eliot's Works*, vol. 22 (Boston: Estes & Lauriat, 1895), 65.

7. Romantics

1. Walter Maunder quoted in R. A. Marriot, "Norway 1896: The BAA's First Organized Eclipse Expedition," *Journal of the British Astronomical Association* 101, no. 3 (1991): 165.
2. Quoted in Willie Wei-Hock Soon and Steven H. Yaskell, *The Maunder Minimum and the Variable Sun-Earth Connection* (River Edge, NJ: World Scientific, 2003), 88.
3. Ibid.
4. Quoted in Alex Soojung-Kim Pang, *Empire and the Sun: Victorian Solar Eclipse Expeditions* (Stanford, CA: Stanford University Press, 2002), 46.
5. Ibid., 58.
6. Stuart Clark, *The Sun Kings: The Unexpected Tragedy of Richard Carrington and the Tale of How Modern Astronomy Began* (Princeton, NJ: Princeton University Press, 2007), 26–27.
7. Stephen Hebron, "The Romantics and Italy," *British Library*, May 15, 2014, www.bl.uk/romantics-and-victorians/articles/the-romantics-and-italy.
8. Percy Bysshe Shelley, "Julian and Maddalo," *Poetry Foundation*, www.poetryfoundation.org/poems/45125/julian-and-maddalo.
9. Clark, *Sun Kings*, 27–28.
10. Pang, *Empire and the Sun*, 76.
11. Ibid., 95.
12. Ibid., 111.
13. Ibid., 72–73.
14. Ibid., 75.
15. Ibid., 74.
16. The remark is repeated in W. Valentine Ball, ed., *Reminiscences and Letters of Sir Robert Ball* (Boston: Little, Brown, 1915), 305.
17. Ibid.
18. Ibid., 312.
19. For details on the coelostat, see J. A. Faterson, "Sir Robert Ball," *Journal of the Royal Astronomical Society of Canada* 10 (1916): 42.
20. Marriot, "Norway 1896," 168.
21. Ibid., 170.
22. Marilyn Bailey Ogilvie, "Obligatory Amateurs: Annie Maunder (1868–1947) and British Women Astronomers at the Dawn of Professional Astronomy," *British Journal of the History of Science* 33, no. 1 (2000): 78–79.

23. W. Valentine Ball, ed., *Reminiscences and Letters*, 316.
24. Mention of the play can be found in P. D. Hingley, "Two Weddings and a Funeral . . . ," *Astronomy and Geophysics* 40 (August 1999): 47.
25. W. Valentine Ball, ed., *Reminiscences and Letters*, 316.
26. E. Walter Maunder, *The Indian Eclipse 1898: Report of the Expeditions Organized by the British Astronomical Association to Observe the Total Eclipse of 1898, January 22* (London: Hazell, Watson & Viney, 1899), 4.
27. Ibid., 5.
28. Ibid.
29. Ibid.
30. Ibid., 9.
31. Ibid., 10.
32. Ibid., 14.
33. Ibid.
34. Ibid., 21.
35. Agnes Clerke cited in Mary Brück, *Women in Early British and Irish Astronomy: Stars and Satellites* (Dordrecht: Springer, 2009), 224.
36. Reynolds, "Introduction to the Second Edition" of Orr, *Dante and the Early Astronomers*, 17.

8. Prisms

1. F. M. Stratton, "John Evershed, 1864–1956," *Biographical Memoirs of Fellows of the Royal Society* 3 (November 1957): 41.
2. Cited in J. B. Zirker, *Total Eclipses of the Sun* (New York: Van Nostrand Reinhold, 1984), 5.
3. N. Kamesara Rao, A. Vagiswari, and Christina Birdie, "Charles Michie Smith—Founder of the Kodaikanal (Solar Physics) Observatory and Beginnings of Physical Astronomy in India," *Current Science* 106, no. 3 (February 10, 2014): 461.
4. Sir Robert Ball told this story on many occasions. It is cited in Faterson, "Sir Robert Ball," 57.
5. Dante, *Inferno*, trans. Hollander and Hollander, 97.

9. The Notebook of the Sun

1. M. K. V. Bappu, "The Kodaikanal Observatory—A Historical Account," *Journal of Astrophysical Astronomy* 21 (2000): 103.

2. Isaac Newton quoted in A. E. Whitford, "Astronomy and Astronomers at Mountain Observatories," *New York Academy of Science* 198 (1972): 202.
3. Rao, Vagiswari, and Birdie, "Charles Michie Smith," 459.
4. Ibid., 460.
5. Ibid.
6. John Evershed, letter to Kate Evershed, August 27, 1922, John Evershed Archive, Science Museum Library and Archives, Science Museum at Wroughton.
7. Nigel Calder, *Einstein's Universe* (New York: Penguin Books, 1979), 24–25.
8. C. Michie Smith, *Kodaikanal Bulletin, 1905* (Madras: Madras Government Publication, 1905). The following quotations are from this document as well.
9. For a discussion of spectroscopic studies of ancient mosaics, see L. de Ferri, D. Bersani, P. H. Columban, P. P. Lottici, G. Simon, and G. Vezdini, "Raman Study of Model Glass with Medieval Compositions: Artificial Weathering and Composition with Ancient Samples," *Journal of Raman Spectroscopy* 43, no. 11 (June 2012).
10. Walter and Annie Maunder cited in Soon and Yaskell, *The Maunder Minimum*, 221.
11. Brandon Brown, e-mail to the author, July 4, 2017.
12. Leo Tolstoy quoted in Carl Gaither and Alma E. Cavazos-Gaither, eds., *Gaither's Dictionary of Scientific Quotations* (New York: Springer Science and Business Media, 2012), 1460.
13. Soon and Yaskell, *The Maunder Minimum*, 2.
14. Ibid., 9. For a detailed discussion of Scheiner, see Galileo Galilei and Christoph Scheiner, *On Sunspots*, translated by Eileen Reeves and Albert Van Helden (Chicago: University of Chicago Press, 2010), 37–57.
15. Soon and Yaskell, *The Maunder Minimum*, 6–7.
16. Ibid., 87.
17. For my summary of the sun's structure I have drawn from the following sources: George Gamow, *The Birth and Death of the Sun* (New York: Penguin Books, 1945); Jean-Claude Pecker, *The Future of the Sun* (New York: McGraw-Hill, 1990); and Soon and Yaskell, *The Maunder Minimum*.
18. J. Evershed, "The Cause of Darkness in Sunspots," *Astrophysical Journal* 5 (1897): 247.
19. J. Evershed, "Observations of Halley's Comet," *Kodaikanal Bulletin*, no. 20 (June 22, 1910): 199–207.

20. John Evershed and Mary Acworth Evershed, *Memoirs of the Kodaikanal Observatory: Results of Prominence Observations* (Madras: Government Press, 1917), 55.
21. M. A. Evershed, "Some Types of Prominences Associated with Sunspots," *Monthly Notices of the Royal Astronomical Society* 73 (1913): 422–430.

10.
The Gift of the Forest

1. For a detailed account of Indian astronomy, see R. K. Kochhar, "The Growth of Modern Astronomy in India," *Vistas in Astronomy* 34 (1991): 69–105. For my historical overview, I have also drawn from the following sources: M. N. Vahia, "Origin and Growth of Astronomy in Indian Context," www.tifr.res.in/~archaeo/FOP/FoP%20papers/growth%20of%20astronomy.pdf; Jayanta Bhadra, "Astronomy, Computer Science, and Mathematics in Ancient India," www.cerc.utexas.edu/~jay/anc.html; Manikant Shah, "Indian Astronomy through Ages," www.infinityfoundation.com/mandala/t_es/t_es_shah_m_astronomy.htm.
2. Vahia, "Origin and Growth," 46.
3. Shah, "Indian Astronomy through Ages."
4. S. Abid Husain, *The National Culture of India* (Bombay: Asia Publishing House, 1961), 26–27.
5. Cited in Mitchell, *Indian Hill-Station: Kodaikanal*, 125.
6. The Kodaikanal Club website, www.thekodaikanalclub.com.
7. Ronald L. Martinez, "Dante between Hope and Despair: The Tradition of Lamentations in the *Divine Comedy*," *Logos* 5, no. 3 (2002): 115.
8. Ibid., 115–117.
9. M. A. Orr, "Astronomy in the Old Testament," *Knowledge and Illustrated Scientific News* 1, no. 27 (October 1904): 234–235. The following quotations are from this article.
10. For a detailed discussion about comets and the Star of Bethlehem, see Colin J. Humphreys, "The Star of Bethlehem—a Comet in 5 BC—and the Date of the Birth of Christ," *Quarterly Journal of the Royal Astronomical Society* 32 (1991): 389–407.

11.
The Scarcity of Wasps in Kashmir

1. Mary Evershed, letter to Kate Evershed, December 12, 1922, John Evershed Archive.

2. C. Michie Smith, *Kodaikanal Bulletin, 1910* (Madras: Madras Government Publication, 1910).
3. Husain, *National Culture of India*, 124. See also Husain's discussion of British-Indian relationships on pp. 122–133 and 134–186.
4. Buettner, "Problematic Spaces, Problematic Races," 278–284.
5. Vinay Lal, "Hill Stations: Pinnacles of the Raj," *The Book Review (Delhi)* 17, no. 9 (September 1993): 8–9.
6. In 1915, the Madras government published a catalog entitled *The Flora of the Nilgiri and Pulney Hilltops*, with contributions from Mary.
7. John Evershed, "A Remarkable Flight of Birds," *Nature* 52, no. 1351 (1895): 508.
8. John Evershed, Letter, *Nature* 81, no. 2235 (August 22, 1912): 659.
9. William Shakespeare, *Troilus and Cressida*, in *The Riverside Shakespeare* (Boston: Houghton Mifflin, 1974), 486.
10. William Shakespeare, *Hamlet*, ibid., 1146.
11. Dante quoted in Orr, *Dante and the Early Astronomers*, 182.
12. Ibid., 288.

12.
Harmonic Structures

1. Edward Moore, "The Astronomy of Dante," in *Studies in Dante, Third Series: Miscellaneous Essays* (Oxford: Oxford University Press, [1903] 1968), 1.
2. Dante, *Inferno*, trans. Musa, 67.
3. See Dante, *Purgatorio*, trans. Musa, 322.
4. Ibid., 345.
5. Moore, "Astronomy of Dante," 55, 106.
6. Orr, *Dante and the Early Astronomers*, 312–316.
7. Ibid., 170–172.
8. Ibid., 174.
9. Ibid.
10. See Dante's *Convivio*, Second Treatise; the fullest translation of the passage can be found in *The Banquet of Dante Alighieri: Il Convivio*, translated by Elizabeth Price (London: George Routledge & Sons, 1887), 85–86.
11. Dante, *Paradiso*, trans. Hollander and Hollander, 307.
12. Ibid., 313.

13. K. C. Cole, *The Hole in the Universe: How Scientists Peered over the Edge of Emptiness and Found Everything* (New York: Harcourt, 2001), 187.
14. See David Castelvecchi, "Dante's Universe, and Ours," *The Nature of Reality*, July 23, 2002, www.pbs.org/wgbh/nova/blogs/physics/2012/07/dantes-universe/.
15. "Cosmic Microwave Background," *Cosmos: The SAO Encyclopedia of Astronomy*, http://astronomy.swin.edu.au/cosmos/C/Cosmic+Microwave+Background.
16. Stephen Hawking. "The Beginning of Time," www.hawking.org.uk/the-beginning-of-time.html.
17. Castelvecchi, "Dante's Universe."
18. Hawking, "Beginning of Time."
19. Castelvecchi, "Dante's Universe."
20. Dante, *Convivio,* Book II, translated by Christopher Ryan as *The Banquet* (Saratoga, CA: ANMA Libri, 1989), 26–27.
21. Dante, *Paradiso*, trans. Hollander and Hollander, 683.
22. Ibid., 663.
23. Ibid.
24. Peterson, *Galileo's Muse*, 74.
25. Grant, *Planets, Stars, and Orbs*, 9–10.
26. Egginton, "Dante, Hyperspheres, and the Curvature of the Medieval Cosmos," 215.

13.
"Dante and the Early Astronomers"

1. Howard L. Kelly, ed., *The History of the British Astronomical Association: The First Fifty Years* (Hounslow West, Middlesex: British Astronomical Association, 1948), 27.
2. Quoted in Orr, *Dante and the Early Astronomers*, 307.
3. Reynolds, "Introduction to the Second Edition," ibid., 20.
4. Virgil, *Georgics*, translated by L. P. Wilkinson (London: Penguin Books, 1982), 71.
5. Orr, *Dante and the Early Astronomers*, 39.
6. Ibid., 168.
7. Ibid., 170.
8. Ibid., 190.
9. Ibid.
10. Ibid., 270.
11. Ibid., 330.

12. Bappu, "The Kodaikanal Observatory," 57–58.
13. Quoted in Brück, "Mary Ackworth Evershed," 54.
14. Orr, *Dante and the Early Astronomers*, 333.
15. Brück, "Mary Ackworth Evershed," 54–55.

14. Sun-Chasers

1. See Mary Brück, "Mountain Paradise," in *Women in Early British and Irish Astronomy*, 235–245.
2. John Evershed cited in Vinayak Razdan, "The Sun Chasers in Kashmir, 1913–14–15–16," *Search Kashmir in Bits and Pieces*, www.searchkashmir.org/2014/08/the-sun-chasers-in-kashmir-1913-14-15-16.html.
3. Galileo and Scheiner, *On Sunspots*, 291.
4. J. Evershed, "A Scarcity of Wasps in Kashmir in 1916" (Letter), *Nature* 99, no. 185 (May 1917).
5. Kelly, ed., *History of the British Astronomical Association*, 29. See also John Earman and Clark Glymour, "Relativity and Eclipses: The British Eclipse Expeditions of 1919 and Their Predecessors," *Historical Studies and the Physical Sciences* 11, no. 1 (1980): 51–52.
6. Judith T. Kenny, "Climate, Race, and Imperial Authority: The Symbolic Landscape of the British Hill Station in India," *Annals of the Association of American Geographers* 85 (1995): 710.
7. For an in-depth discussion of the Bastar uprising, see Ajay Verghese, "British Rule and Tribal Revolts in India: The Curious Case of Bastar," *Modern Asian Studies* 50, no. 5 (September 2016), www.cambridge.org/core/journals/modern-asian-studies/article/british-rule-and-tribal-revolts-in-india-the-curious-case-of-bastar/D1F52864F0CBD6D37FFC2B7DE9CBA5A4/core-reader.
8. See B. S. Baliga, *Studies in Madras Administration*, vol. 1 (Madras: Printed at the India Press, 1960), 38; K. Ragayayan, *History of Tamil Nadu 1565–1982* (Madurai: Raj, 1982): 182; and A. R. Venkatachalapathy, *Economic and Political Weekly* 45, no. 2 (January 9, 2010), 37. For a broader overview of the Raj and Indian independence, see Denis Judd, *The Lion and the Tiger: The Rise and Fall of the British Raj* (Oxford: Oxford University Press, 2004).
9. John Evershed, *Kodaikanal Bulletin, 1916* (Madras: Madras Government Publication, 1916).

15.
Exploding the Sun

1. Milton quoted in Fermi and Bernardini, *Galileo and the Scientific Revolution*, 95.
2. Dante, *Paradiso*, trans. Hollander and Hollander, 235.
3. John Milton, *Paradise Lost*, edited by J. J. Shawcross and Michael Lieb (Pittsburgh, PA: Duquesne University Press, 2007), 591.
4. See Fermi and Bernardini, *Galileo and the Scientific Revolution*, 111.
5. Ibid.
6. See Matthew J. Parry-Hill and Michael W. Davidson, "Thomas Young's Double Slit Experiment," *Molecular Expressions: Optical Microscopy Primer—Physics of Light and Color*, http://micro.magnet.fsu.edu/primer/java/interference/doubleslit.
7. See Charles Paul Enz, *No Time to Be Brief: A Scientific Biography of Wolfgang Pauli* (Oxford: Oxford University Press, 2002), 388.
8. Details of Einstein's experiments are adapted from David Tong, "Einstein and Relativity: Part I," *Plus*, http://plus.maths.org/content/einstein-relativity; and from Calder, *Einstein's Universe*, 11.
9. Calder, *Einstein's Universe*, 166.

16.
Saturnalia

1. Kelly, ed., *History of the British Astronomical Association*, 29.
2. D. H. Lawrence, letter to Lady Ottoline Morrell, September 9, 1915, in *Selected Letters of D. H. Lawrence*, edited by James Boulton (Cambridge: Cambridge University Press, 2000), 106.
3. Kelly, ed., *History of the British Astronomical Association*, 24.
4. John Evershed, *Kodaikanal Bulletin, 1919* (Madras: Madras Government Publication, 1919).
5. John Evershed, *Kodaikanal Bulletin, 1921* (Madras: Madras Government Publication, 1921).
6. Quoted in Philip Ball, "How 2 Pro-Nazi Nobelists Attacked Einstein's 'Jewish Science,'" *Scientific American*, February 13, 2015, www.scientificamerican.com/article/how-2-pro-nazi-nobelists-attacked-einstein-s-jewish-science.

17.
Infinity and the Fly

1. Quoted in Matthew Stanley, "An Expedition to Heal the Wounds of War: The 1919 Eclipse and Eddington as Quaker Adventurer," *ISIS* 94, no. 1 (2003): 57, 65.
2. Eddington quoted in Daniel Kennefick, "Not Only because of Theory: Dyson, Eddington, and the Competing Myths of the 1919 Eclipse Expedition," *Einstein Studies*, November 12, 2011, 201.
3. John Evershed, "The Einstein Effect and the Eclipse of 1919 May 29," *Observatory* 515 (1917): 269–270.
4. For a broad overview of the 1919 solar eclipse expeditions, see Jeffrey Crelinsten, *Einstein's Jury: The Race to Test Relativity* (Princeton, NJ: Princeton University Press, 2006).
5. See the letter from Jay M. Passachoff, director of the Hopkins Observatory at Williams College, in *Science* 317 (October 12, 2007): 210–211.
6. Alistair Sponsel, "Constructing a 'Revolution in Science': The Campaign to Promote a Favourable Reception for the 1919 Solar Eclipse Experiments," *British Journal for the History of Science* 35, no. 4 (December 2002): 461.
7. Ibid., 441, for this and the following newspaper headlines.
8. Earman and Glymour, "Relativity and Eclipses," 83.
9. Rutherford cited in Stanley, "An Expedition to Heal the Wounds of War," 58.
10. Sponsel, "Constructing a 'Revolution in Science,'" 448.
11. Ibid., 462–463.
12. Ibid.
13. Ibid., 465.
14. Ibid., 462.
15. Ibid.

18.
Wallal

1. Mary Evershed discussing John's idea in a letter to Kate Evershed, January 29, 1923, John Evershed Archive.
2. See Eric Gray Forber, "A History of the Solar Redshift Problem," *Annals of Science* 17, no. 3 (December 1963): 137.

3. Dr. Edwin Slosson, "The Progress of Science," *Scientific Monthly* 15, no. 1 (July 1922): 475.
4. This and all other quotes pertaining to the Evershed's' Wallal expedition are from J. Evershed, "Report of the West Indian Eclipse Expedition to Wallal, West Australia," *Kodaikanal Observatory Bulletin*, no. 72, 1923. Further details are drawn from C. A. Chant, "The Eclipse Camp at Wallal," *Journal of the Royal Astronomical Society of Canada* 17, no. 1 (January–February 1923).
5. Details are from John Evershed, letter to Kate Evershed, November 20, 1923, John Evershed Archive.
6. RAS report cited at http://australia-tests-einstein.weebly.com/about.html.
7. Campbell, "Total Eclipse of the Sun," 36.
8. John Evershed to Kate Evershed, November 20, 1923.
9. Ibid.
10. W. W. Campbell, "The Total Eclipse of the Sun, September 21, 1922," *Astronomical Society of the Pacific* 35 (1923): 36.
11. John Evershed, *Kodaikanal Bulletin, 1922* (Madras: Madras Government Publication, 1922). Subsequent quotations are from this report unless otherwise noted.
12. Alexander David Ross quoted in John C. Robins, *History of the Department of Physics at UWA*, no. 9: "Wallal: The 1922 Solar Eclipse Expedition to Test Einstein's Theory," 8.
13. John Evershed, letter to Kate Evershed, February 26, 1923, John Evershed Archive.
14. *New York Times* quoted in Lisa Grossman, "Can the Eclipse Tell Us If Einstein Was Right about General Relativity?" *Science News*, August 15, 2017, www.sciencenews.org/article/2017-solar-eclipse-einstein-general-relativity.

19. Departure

1. Mary Evershed, letter to Kate Evershed, December 5, 1922, John Evershed Archive.
2. The following account is shared in a letter from Mary Evershed to Kate Evershed, February 18, 1923, ibid.
3. Mary Evershed, letter to Kate Evershed, January 29, 1923, ibid.
4. John Evershed, letter to Kate Evershed, February 26, 1923, ibid.
5. Mary Evershed, letter to Kate Evershed, February 18, 1923.
6. John Evershed, letter to Kate Evershed, February 26, 1923.

20.
Who's Who in the Moon

1. Kelly, ed., *History of the British Astronomical Association*, 126.
2. Ibid., 33.
3. Mary Evershed, letter to Harry Evershed, April 8, 1923, John Evershed Archive.
4. John Evershed, letter to Harry Evershed, April 3, 1923, ibid.
5. For an overview of recent theories about Homer's eclipse reference, see J. R. Minkel, "Homer's *Odyssey* Said to Document 3,200-Year-Old Eclipse," *Scientific American*, June 23, 2008, www.scientificamerican.com/article/homers-odyssey-may-document-eclipse.
6. Dante, *Purgatorio*, trans. Hollander and Hollander, 327.
7. Aristotle, *On the Generation of Animals*, Book I, translated by Arthur Platt in *The Oxford Translation of Aristotle*, edited by W. D. Ross (Oxford: Clarendon Press, 1912), unpaginated.
8. Woolf, *To the Lighthouse*, 129.
9. Carlyle and Mary Evershed quoted in Kelly, ed., *History of the British Astronomical Association*, 128.
10. Ibid., 127–128.
11. M. A. Evershed, Introduction to *Who's Who in the Moon*, Memoirs of the British Astronomical Association, vol. 34 (London, 1938), 1.
12. *Who's Who in the Moon* entries cited in "Review of Publications," *Journal of the Royal Astronomical Society of Canada* 33 (1939): 67–68.
13. Ibid., 67.

21.
The Maunder Minimum

1. Cited in Klaus Hentschel, "The Conversion of St. John: A Case Study of the Interplay of Theory and Experiment," *Science in Context* 6, no. 1 (1993): 164–165.
2. Ibid.
3. Soon and Yaskell, *The Maunder Minimum*, 141, 26.
4. Walter Maunder quoted ibid., 233.
5. A. E. Douglass, *Climatic Cycles and Tree Growth: A Study of the Annual Rings of Trees in Relation to Climate and Solar Activity* (Washington, DC: Carnegie Institution of Washington, 1919).
6. Soon and Yaskell, *The Maunder Minimum*, 134–145.
7. Ibid., 26.

8. Andrew Marvell, "The Last Instructions to a Painter," quoted in J. E. Weiss and N. O. Weiss, "Andrew Marvell and the Maunder Minimum," *Quarterly Journal of the Royal Astronomical Society* 20 (1979): 116.
9. Shakespeare, *The Tempest*, in *The Riverside Shakespeare*, 1622.
10. Quoted in Soon and Yaskell, *The Maunder Minimum*, 58.
11. Jonathan Swift, *Gulliver's Travels* (Boston: Riverside Press, 1960), 132.
12. William Herschel cited in Soon and Yaskell, *The Maunder Minimum*, 74.
13. Quoted ibid., 125.

22.
The Remade Universe

1. Kelly, ed., *History of the British Astronomical Association*, 35.
2. Ibid., 47.
3. Mary Evershed, Introduction to T. E. R. Phillips, ibid., 200.
4. T. E. R. Phillips, "Some Recent Views of the Physical Universe and Their Reaction on Present Day Thought," ibid., 183–184.
5. Ibid., 184.
6. Ibid., 196–200.
7. Ibid., 201–205.

23.
Return to Origins

1. M. A. Evershed, "Arab Astronomy," *Observatory* 58 (August 1935): 243.
2. Ibid., 238.
3. Ibid., 237.

24.
Northern Lights

1. See Stratton, "John Evershed, 1864–1956."
2. Quoted in Reynolds, "Introduction to the Second Edition" of Orr, *Dante and the Early Astronomers*, 19.
3. Quoted in Brück, "Mary Ackworth Evershed," 56.
4. Dante, *De Vulgari Eloquentia*, Book I, Chapter VI. For the fullest translation of the passage, see Marianne Shapiro, *De Vulgari*

Eloquentia: Dante's Book of Exile (Lincoln: University of Nebraska Press, 1990), 65.

5. Barbara Reynolds, *The Passionate Intellect: Dorothy L. Sayers' Encounters with Dante* (Kent, OH: Kent State University Press, 1989), 117, 225.
6. Colin Hardie, Review of *Dante and the Early Astronomers*, *Modern Language Review* 52, no. 4 (October 1957): 614.
7. Reynolds, "Introduction to the Second Edition" of Orr, *Dante and the Early Astronomers*, 20.
8. Albert Einstein quoted in Rasoul Sorkhabi, "Einstein and Indian Minds: Tagore, Gandhi, and Nehru," *Current Science* 88, no. 7 (April 10, 2005): 1189.
9. Dante, *Paradiso*, trans. Hollander and Hollander, 3.
10. Harold W. Newton, letter to John Evershed, August 25, 1949, John Evershed Archive.
11. See John Evershed's summary of his career, "Recollections of Seventy Years of Scientific Work," *Vistas in Astronomy* 1, no. 33 (1955): 33–40.
12. Stratton, "John Evershed, 1864–1956," 48.
13. J. Evershed, "Recollections of Seventy Years of Scientific Work," 36, 38.
14. Dante, *Paradiso*, trans. Hollander and Hollander, 767.
15. Peter S. Hawkins, *Dante: A Brief History* (Malden, MA: Blackwell, 2006), 81.

Epilogue: Kodai Dusk

1. Evershed and Evershed, *Memoirs of the Kodaikanal Observatory*, n.p.
2. Dante, *Paradiso*, trans. Hollander and Hollander, 819–820.
3. Bhavita B, "Keepers of the Sun—Untold," Indian Institute of Astrophysics and Kodiakanal Solar Observatory, www.untold.in/keepers-of-the-sun. See also Reuters, "At Indian Observatory, Family Records Daily Life of the Sun," February 23, 2017, *VOA*, www.voanews.com/a/indian-observatory-records-daily-activity-of-sun/3736339.html.
4. Ibid.
5. Peterson, "Dante and the 3-Sphere," 1031.
6. Cornish, *Reading Dante's Stars*, 10.

Bibliography

Aristotle. *On the Generation of Animals*, Book I. Translated by Arthur Platt in *The Oxford Translation of Aristotle*, edited by W. D. Ross. Oxford: Clarendon Press, 1912.

B, Bhavita. "Keepers of the Sun—Untold." Indian Institute of Astrophysics and Kodiakanal Solar Observatory. www.untold.in/keepers-of-the-sun.

Baliga, B. S. *Studies in Madras Administration*. Vol. 1. Madras: Printed at the India Press, 1960.

Ball, Philip. "How 2 Pro-Nazi Nobelists Attacked Einstein's 'Jewish Science.'" *Scientific American*, February 13, 2015, www.scientificamerican.com/article/how-2-pro-nazi-nobelists-attacked-einstein-s-jewish-science.

Ball, W. Valentine, ed. *Reminiscences and Letters of Sir Robert Ball*. Boston: Little, Brown, 1915.

Bappu, M. K. V. "The Kodaikanal Observatory—A Historical Account." *Journal of Astrophysical Astronomy* 21 (2000).

Bergin, Thomas G. *Dante*. New York: Orion, 1965.

Birk, Sandow, and Marcus Sanders. *Dante's Inferno*. San Francisco: Chronicle Books, 2004.

Boccaccio. *Life of Dante*. Translated by J. G. Nichols. London: Hesperus, 2002.

Brück, Mary. *Agnes Mary Clerke and the Rise of Astrophysics*. Cambridge: Cambridge University Press, 2000.

———. "Mary Ackworth Evershed née Orr (1867–1949), Solar Physicist and Dante Scholar." *Journal of Astronomical History and Heritage* 1, no. 1 (1998).

———. *Women in Early British and Irish Astronomy: Stars and Satellites*. Dordrecht: Springer, 2009.

Buettner, Elizabeth. "Problematic Spaces, Problematic Races: Defining 'Europeans' in Late Colonial India." *Women's History Review* 9, no. 2 (2000).

Calder, Nigel. *Einstein's Universe*. New York: Penguin Books, 1979.

Campbell, W. W. "The Total Eclipse of the Sun, September 21, 1922." *Astronomical Society of the Pacific* 35 (1923).

Castelvecchi, David. "Dante's Universe, and Ours." *The Nature of Reality,* July 23, 2002, www.pbs.org/wgbh/nova/blogs/physics/2012/07/dantes-universe.

Chant, C. A. "The Eclipse Camp at Wallal." *Journal of the Royal Astronomical Society of Canada* 17, no. 1 (January–February 1923).

Chubb, Thomas Caldecott. *Dante and His World.* New York: Little, Brown, 1966.

Clark, Stuart. *The Sun Kings: The Unexpected Tragedy of Richard Carrington and the Tale of How Modern Astronomy Began.* Princeton, NJ: Princeton University Press, 2007.

Cole, K. C. *The Hole in the Universe: How Scientists Peered over the Edge of Emptiness and Found Everything.* New York: Harcourt, 2001.

Connor, Steve. "The Core of Truth behind Sir Isaac Newton's Apple." *Independent,* January 18, 2010, independent.co.uk/news/science/the-core-of-truth-behind-sir-isaac-newtons-apple-1870915.html.

Cornish, Alison. *Reading Dante's Stars.* New Haven, CT: Yale University Press, 2000.

Crelinsten, Jeffrey. *Einstein's Jury: The Race to Test Relativity.* Princeton, NJ: Princeton University Press, 2006.

Dante Alighieri. *The Banquet of Dante Alighieri: Il Convivio.* Translated by Elizabeth Price. London: George Routledge & Sons, 1887.

———. *The Banquet.* Translated by Christopher Ryan. Saratoga, CA: ANMA Libri, 1989.

———. *De Vulgari Eloquentia: Dante's Book of Exile.* Translated by Marianne Shapiro. Lincoln: University of Nebraska Press, 1990.

———. *Inferno.* Translated by Mark Musa. New York: Penguin Books, 1984.

———. *Inferno.* Translated by Jean Hollander and Robert Hollander. New York: Anchor Books, 2000.

———. *La Vita Nuova.* Translated by Barbara Reynolds. London: Penguin Books, 2004.

———. *Paradiso.* Translated by Mark Musa. New York: Penguin Books, 1986.

———. *Paradiso.* Translated by Jean Hollander and Robert Hollander. New York: Doubleday, 2007.

———. *Purgatorio.* Translated by Mark Musa. New York: Penguin Books, 1985.

———. *Purgatorio.* Translated by Jean Hollander and Robert Hollander. New York: Anchor Books, 2003.

———. *A Translation of the Latin Works of Dante Alighieri.* Translated by A. G. Ferrers Howell and Philip H. Wicksteed. London: J. M. Dent, 1904.

Earman, John, and Clark Glymour. "Relativity and Eclipses: The British Eclipse Expeditions of 1919 and Their Predecessors." *Historical Studies and the Physical Sciences* 11, no. 1 (1980).

Egginton, William. "Dante, Hyperspheres, and the Curvature of the Medieval Cosmos." *Journal of the History of Ideas* 60, no. 2 (April 1999).

Eliot, T. S. *The Sacred Wood.* London: Methuen, 1920.

Enz, Charles Paul. *No Time to Be Brief: A Scientific Biography of Wolfgang Pauli.* Oxford: Oxford University Press, 2002.

Evershed, John. "The Cause of Darkness in Sunspots." *Astrophysical Journal* 5 (1897).

———. "The Einstein Effect and the Eclipse of 1919 May 29." *Observatory* 515 (1917).

———. *Kodaikanal Bulletin, 1916.* Madras: Madras Government Publications, 1916.

———. *Kodaikanal Bulletin, 1919.* Madras: Madras Government Publications, 1919.

———. *Kodaikanal Bulletin, 1921.* Madras: Madras Government Publications, 1921.

———. *Kodaikanal Bulletin, 1922.* Madras: Madras Government Publications, 1922.

———. *Kodaikanal Bulletin, 1923.* Madras: Madras Government Publications, 1923.

———. Letter. *Nature* 81, no. 2235 (August 22, 1912).

———. "Observations of Halley's Comet." *Kodaikanal Bulletin*, no. 20 (June 22, 1910).

———. "Recollections of Seventy Years of Scientific Work." *Vistas in Astronomy* 1, no. 33 (1955).

———. "A Remarkable Flight of Birds." *Nature* 52, no. 1351 (1895).

———. "Report of the West Indian Eclipse Expedition to Wallal, West Australia." *Kodaikanal Observatory Bulletin*, no. 72 (1923).

———. "A Scarcity of Wasps in Kashmir in 1916" (Letter). *Nature* 99, no. 185 (May 1917).

Evershed, John, and Mary Acworth Evershed. *Memoirs of the Kodaikanal Observatory: Results of Prominence Observations.* Madras: Government Press, 1917.

Evershed, M. A. "Arab Astronomy." *Observatory* 58 (August 1935).

———. "Some Types of Prominences Associated with Sunspots." *Monthly Notices of the Royal Astronomical Society* 73 (1913).
———. *Who's Who in the Moon*. Memoirs of the British Astronomical Association, vol. 34. London, 1938.
Evershed, M. A., and J. Evershed. "Dante and Medieval Astronomy." *Observatory* 34 (1911).
Faterson, J. A. "Sir Robert Ball" in the *Journal of the Royal Astronomical Society of Canada*, Volume 10 (1916).
Fermi, Laura, and Gilberto Bernardini. *Galileo and the Scientific Revolution*. Mineola, NY: Dover, 2003.
Forber, Eric Gray. "A History of the Solar Redshift Problem." *Annals of Science* 17, no. 3 (December 1963).
Fyson, P. F., and Lady Bourne. *The Flora of the Nilgiri and Pulney Hill-tops*. Vol. 1. Madras: Madras Government Publications, 1915.
Galileo Galilei. *Two Lectures to the Florentine Academy on the Shape, Location, and Size of Dante's "Inferno"* (1588). Translated by Mark A. Peterson. www.mtholyoke.edu/courses/mpeterso/galileo/inferno.html.
Galileo Galilei and Christoph Scheiner. *On Sunspots*. Translated by Eileen Reeves and Albert Van Helden. Chicago: University of Chicago Press, 2010.
Gamow, George. *The Birth and Death of the Sun*. New York: Penguin Books, 1945.
Godfrey, Barry. "The Australian Colonies, 1787–1901." *The Digital Panopticon Project*, www.digitalpanopticon.org/The_Australian_Colonies,_1787–1901.
Grant, Edward. *Much Ado about Nothing: Theories of Space and Vacuum from the Middle Ages to the Scientific Revolution*. Cambridge: Cambridge University Press, 1981.
———. *Physical Science in the Middle Ages*. Cambridge: Cambridge University Press, 1977.
———. *Planets, Stars, and Orbs: The Medieval Cosmos, 1200–1687*. Cambridge: Cambridge University Press, 1994.
Grossman, Lisa. "Can the Eclipse Tell Us If Einstein Was Right about General Relativity?" *Science News*, August 15, 2017, www.sciencenews.org/article/2017-solar-eclipse-einstein-general-relativity.
Hardie, Colin. Review of *Dante and the Early Astronomers*. *Modern Language Review* 52, no. 4 (October 1957).
Hawking, Stephen. "The Beginning of Time." www.hawking.org.uk/the-beginning-of-time.html.
Hawkins, Peter S. *Dante: A Brief History*. Malden, MA: Blackwell, 2006.

Hawkins, Peter S., and Rachel Jacoff, eds. *The Poet's Dante: Twentieth-Century Responses*. New York: Farrar, Straus & Giroux, 2001.

Hebron, Stephen. "The Romantics and Italy." *British Library*, May 15, 2014, www.bl.uk/romantics-and-victorians/articles/the-romantics-and-italy.

Hentschel, Klaus. "The Conversion of St. John: A Case Study of the Interplay of Theory and Experiment." *Science in Context* 6, no. 1 (1993).

Hingley, P. D. "Two Weddings and a Funeral . . ." *Astronomy and Geophysics* 40 (August 1999).

Humphreys, Colin J. "The Star of Bethlehem—a Comet in 5 BC—and the Date of the Birth of Christ." *Quarterly Journal of the Royal Astronomical Society* 32 (1991).

Husain, S. Abid. *The National Culture of India*. Bombay: Asia Publishing House, 1961.

Judd, Denis. *The Lion and the Tiger: The Rise and Fall of the British Raj*. Oxford: Oxford University Press, 2004.

Kelly, Howard L., ed. *The History of the British Astronomical Association: The First Fifty Years*. Hounslow West, Middlesex: British Astronomical Association, 1948.

Kennefick, Daniel. "Not Only because of Theory: Dyson, Eddington, and the Competing Myths of the 1919 Eclipse Expedition." *Einstein Studies*, November 12, 2011.

Kenny, Judith T. "Climate, Race, and Imperial Authority: The Symbolic Landscape of the British Hill Station in India." *Annals of the Association of American Geographers* 85 (1995).

Kochhar, R. K. "The Growth of Modern Astronomy in India." *Vistas in Astronomy* 34 (1991).

Lal, Vinay. "Hill Stations: Pinnacles of the Raj." *The Book Review* (*Delhi*) 17, no. 9 (September 1993).

Lawrence, D. H. *Selected Letters of D. H. Lawrence*. Edited by James Boulton. Cambridge: Cambridge University Press, 2000.

Marriot, R. A. "Norway 1896: The BAA's First Organized Eclipse Expedition." *Journal of the British Astronomical Association* 101, no. 3 (1991).

Martinez, Ronald L. "Dante between Hope and Despair: The Tradition of Lamentations in the *Divine Comedy*." *Logos* 5, no. 3 (2002).

Maunder, E. Walter. *The Indian Eclipse 1898: Report of the Expeditions Organized by the British Astronomical Association to Observe the Total Eclipse of 1898, January 22*. London: Hazell, Watson & Viney, 1899.

Milton, John. *Paradise Lost*. Edited by J. J. Shawcross and Michael Lieb. Pittsburgh, PA: Duquesne University Press, 2007.

Minkel, J. R. "Homer's *Odyssey* Said to Document 3,200-Year-Old Eclipse." *Scientific American*, June 23, 2008.

Mitchell, Nora. *The Indian Hill Station: Kodaikanal*. Chicago: University of Chicago, Department of Geography, 1972.

Moore, Edward. "The Astronomy of Dante." In *Studies in Dante, Third Series: Miscellaneous Essays*. Oxford: Oxford University Press, [1903] 1968).

Ogilvie, Marilyn Bailey. "Obligatory Amateurs: Annie Maunder (1868–1947) and British Women Astronomers at the Dawn of Professional Astronomy." *British Journal of the History of Science* 33, no. 1 (2000).

Orchiston, Wayne. *John Tebbutt: Rebuilding and Strengthening the Foundations of Australian Astronomy*. Cham, Switz.: Springer International, 2016.

Orr, M. A. "Astronomy in the Old Testament." *Knowledge and Illustrated Scientific News* 1, no. 27 (October 1904).

———. "Black Star-Lore," *Journal of the British Astronomical Association* 9, no. 680 (1898).

———. *Dante and the Early Astronomers*. Port Washington, NY: Kennikat Press, [1913] 1969.

———. *An Easy Guide to the Southern Stars*. London and Edinburgh: Gall & Inglis, 1897.

Pang, Alex Soojung-Kim. *Empire and the Sun: Victorian Solar Eclipse Expeditions*. Stanford, CA: Stanford University Press, 2002.

Parry-Hill, Matthew, and Michael W. Davidson. "Thomas Young's Double Slit Experiment." *Molecular Expressions: Optical Microscopy Primer—Physics of Light and Color*, http://micro.magnet.fsu.edu/primer/java/interference/doubleslit.

Pecker, Jean-Claude. *The Future of the Sun*. New York: McGraw-Hill, 1990.

Peterson, Mark A. "Dante and the 3-Sphere." *American Journal of Physics* 47, no. 12 (December 1979).

———. *Galileo's Muse: Renaissance Mathematics and the Arts*. Cambridge, MA: Harvard University Press, 2011.

Pollack, Eileen. *The Only Woman in the Room: Why Science Is Still a Boys' Club*. Boston: Beacon Press, 2015.

Ragayayan, K. *History of Tamil Nadu 1565–1982*. Madurai: Raj, 1982.

Rao, N. Kamesara, A. Vagiswari, and Christina Birdie. "Charles Michie Smith—Founder of the Kodaikanal (Solar Physics) Observatory and

Beginnings of Physical Astronomy in India." *Current Science* 106, no. 3 (February 10, 2014).
Razdan, Vinayak. "The Sun Chasers in Kashmir, 1913–14–15–16." *Search Kashmir in Bits and Pieces*, www.searchkashmir.org/2014/08/the-sun-chasers-in-kashmir-1913-14-15-16.html.
Reeves, Eileen. "From Dante's Moonspots to Galileo's Sunspots." *MLN* 124, no. 5 (2009).
Reynolds, Barbara. *The Passionate Intellect: Dorothy L. Sayers' Encounters with Dante*. Kent, OH: Kent State University Press, 1989.
Robins, John C. *History of the Department of Physics at UWA*, no. 9: "Wallal: The 1922 Solar Eclipse Expedition to Test Einstein's Theory."
Rubin, Harriet. *Dante in Love: The World's Greatest Poem and How It Made History*. New York: Simon & Schuster, 2004.
Ruskin, John. *The Works of John Ruskin*, vol. 23. Edited by E. T. Cook and Alexander Wedderburn. London: George Allen, 1906.
Shah, Manikant. "Indian Astronomy Through Ages." www.infinityfoundation.com/mandala/t_es/t_es_shah_m_astronomy.htm.
Shakespeare, William. *The Riverside Shakespeare*. Boston: Houghton Mifflin, 1974.
Siddiqui, Danish. "Kodaikanal's Sungazers." Reuters, February 23, 2017, https://widerimage.reuters.com/story/kodaikanals-sungazers.
Smith, C. Michie. *Kodaikanal Bulletin, 1905*. Madras: Madras Government Publications, 1905.
———. *Kodaikanal Bulletin, 1910*. Madras: Madras Government Publications, 1910.
Sobel, Dava. *The Glass Universe: How the Ladies of the Harvard Observatory Took the Measure of the Stars*. New York: Viking, 2016.
Soon, Willie Wei-Hock, and Steven H. Yaskell. *The Maunder Minimum and the Variable Sun-Earth Connection*. River Edge, NJ: World Scientific, 2003.
Sorkhabi, Rasoul. "Einstein and Indian Minds: Tagore, Gandhi, and Nehru." *Current Science* 88, no. 7 (April 10, 2005).
Soundarapandian, Mookkiah. *Development of Special Economic Zones in India: Policies and Issues*. Delhi: Concept, 2012.
Sponsel, Alistair. "Constructing a 'Revolution in Science': The Campaign to Promote a Favourable Reception for the 1919 Solar Eclipse Experiments." *British Journal of the History of Science* 35, no. 4 (December 2002).
Stanley, Matthew. "An Expedition to Heal the Wounds of War: The 1919 Eclipse and Eddington as Quaker Adventurer." *ISIS* 94, no. 1 (2003).

Stratton, F. M. "John Evershed, 1864–1956." *Biographical Memoirs of Fellows of the Royal Society* 3 (November 1957).
Swift, Jonathan. *Gulliver's Travels*. Boston: Riverside Press, 1960.
Thomas Aquinas. *Introduction to Saint Thomas Aquinas: The Essence of the "Summa Theologica" and the "Summa Contra Gentiles."* Translated and edited by Anton Pegis. New York: Modern Library, 1948.
Vahia, M. N. "Origin and Growth of Astronomy in Indian Context." www.tifr.res.in/~archaeo/FOP/FoP%20papers/growth%20of%20astronomy.pdf.
Valleriania, Matteo. "An Organ Pipe as a Telescope." *Max-Planck-Gesellschaft*: www.mpg.de/7913340/Galileo¬_Galilei_telescope.
Venkatachalapathy, A. R. [No title.] *Economic and Political Weekly* 45, no. 2 (January 9, 2010).
Verghese, Ajay. "British Rule and Tribal Revolts in India: The Curious Case of Bastar." *Modern Asian Studies* 50, no. 5 (September 2016), www.cambridge.org/core/journals/modern-asian-studies/article/british-rule-and-tribal-revolts-in-india-the-curious-case-of-bastar/D1F52864F0CBD6D37FFC2B7DE9CBA5A4/core-reader.
Virgil. *Georgics*. Translated by L. P. Wilkinson. London: Penguin Books, 1982.
Weiss, J. E., and N. O. Weiss. "Andrew Marvell and the Maunder Minimum." *Quarterly Journal of the Royal Astronomical Society* 20 (1979).
Whitford, A. E. "Astronomy and Astronomers at Mountain Observatories." *New York Academy of Science* 198 (1972).
Wilkens, Ernest Hatch. "Dante and the Mosaics of the Bel San Giovanni." *Speculum* 2, no. 1 (January 1927).
Woolf, Virginia. *To the Lighthouse.* Orlando, FL: Harcourt, [1927] 2005.
Zirker, J. B. *Total Eclipses of the Sun*. New York: Van Nostrand Reinhold, 1984.

Index